CULTIVATING GRASS-ROOTS FOR REGIONAL DEVELOPMENT IN A GLOBALISING ECONOMY

T0144150

In fond memory of Nellie and Minnie

and when Irish eyes were smiling

Cultivating Grass-Roots for Regional Development in a Globalising Economy

Innovation and entrepreneurship in organised markets

JAMES CÉCORA

Routledge
Taylor & Francis Group

LONDON AND NEW YORK

First published 1999 by Ashgate Publishing

Reissued 2018 by Routledge
2 Park Square, Milton Park, Abingdon, Oxon OX14 4RN
711 Third Avenue, New York, NY 10017, USA

Routledge is an imprint of the Taylor & Francis Group, an informa business

Publisher's Note
The publisher has gone to great lengths to ensure the quality of this reprint but points out that some imperfections in the original copies may be apparent.

Disclaimer
The publisher has made every effort to trace copyright holders and welcomes correspondence from those they have been unable to contact.

A Library of Congress record exists under LC control number: 98074705

ISBN 13: 978-1-138-61227-3 (hbk)
ISBN 13: 978-1-138-61229-7 (pbk)
ISBN 13: 978-0-429-46349-5 (ebk)

Contents

About the Author

Dr. James Cécora, was born on June 11, 1944, in Bronxville, N.Y. After finishing grammar and high school, he received a B.A. in liberal arts at Iona College, New Rochelle, N.Y. A military service in Vietnam followed. In 1969, Cécora came to Europe. After taking courses in German language, sociology and journalism at the University of Munich (Germany), he studied urbanism and demography at the University of Paris (France) 1971-73, ending up with a *licences dès sciences sociales,* followed by additional interdisciplinary training in urban and regional planning 1974-76 at the Technical University of Karlsruhe (Germany). He then started working as a research associate at the Fraunhofer Institute of Systems Technology and Innovation Research at Karlsruhe. In 1979, Cécora became a research fellow at the Institute of Structural Research of the German Federal Agricultural Research Centre at Braunschweig.

Here he joined a newly created work group, dedicated to the study and evaluation of living conditions in rural areas. The implementation of time and material resources in rural family households; the income derived from these sources; and the use of this income was a major theme in Cecora's research. With original research ideas and methods, and innovative projects which served as the basis for ongoing areas of research, Cécora was also noted for his enthusiastic participation in empirical studies and evaluation. In 1985 he received his doctorate in home economics from the Justus Liebig University in Giessen with a dissertation on the topic *The Influence of Settlement Structures on the Lifestyle of Private Households.* Numerous publications in the years following are testimony to his broadening of this topic.

From the very beginning, Dr. Cécora sought intensive contacts to colleagues in neighbouring countries and abroad, motivated by the desire for scientific exchange and intellectual stimulation, and assisted by a phenomenal ability to speak foreign languages. A significant result of these contacts was the initiation of his own international and interdisciplinary work groups, which convened several times in the 1990's to examine the behaviour of rural family households, their value systems and opinions as well as social relationships within various cultural, economic and political frameworks. More recently he turned his attention to the impact of innovative and dynamic entrepreneurs on the development of rural areas, as well as the personality characteristics and cultural, social and familial circumstances of such entrepreneurial individuals. The results of his investigations into this topic found entrance into a manuscript on which the book presented here is based.

On June 15, 1998, Dr. James Cécora died unexpectedly. He left a wife and three children.

Introduction

Bottom-up strategies for sustainable regional development[1] must ultimately target specific individuals, groups of individuals, or institutions in the region. This is not to suggest virtue in 'muddling through', case by case, without comprehensive concepts. Knowledge of general regularities in socio-economic interaction are essential for devising efficient and equitable means of allocating public resources to promote regional development, i.e. for formulating and implementing effective development policies. The point is, we are dealing with individual, multidimensional social actors and not simply with atomised, utility-optimising, and interchangeable economic agents. Of course, development policy surely cannot be tailored to individual personages. Smilor and Feeser[2] concluded that public policies attempting to target individual firms for support on the basis of judgements that they be 'winners' are destined to failure due to a high degree of unpredictability of firm survival. Nevertheless, we must try to better comprehend various dimensions exerting significant influences on the individual's economic and social behaviour. This contribution takes a strong position on a number of vital issues pertaining to the need to reorient priorities of economic development policies in accordance with the current, and widely misunderstood socio-economic framework of 'globalisation'. When considered to be initiators, major proponents, or classic representatives of ideas or concepts on which our theoretical frame-work is founded, authors are cited regardless of the date of publication.

Chapter One explains the significance of endogenous and innovative entrepreneurship (a region's prime 'human capital') for sustainable regional development and depicts basic socio-institutional and political problems in promoting it: policy problems, such as inability of programs guided by mainstream, neoclassical economic concepts to adequately deal with regional and local economic phenomena strongly determined by 'non-economic' factors. Implementation problems, such as overregulation of economic activity, rewards to 'dependency culture', and domination of resource flows by entrenched and obsolete structures, as is the case of 'rural development' dominated by agricultural institutions and interest groups, self-confined within their own 'rural-agricultural cosmos'.

These problems are compounded by globalisation processes described in Chapter Two. All signs point to a rapid and uncontrolled amassing of wealth and power by international corporate and financial oligarchies which have, to a great extent and with the help of neoliberal economic policies, emancipated themselves from national and democratic control but which have succeeded in superceding market forces by organising and controlling markets themselves. Populations in formerly socialist countries may be astounded to observe that national centrally planned economies are gradually giving way to a global, increasingly centrally planned economy. In view of recognisable trends increasing socio-economic disparities between regions induced by globalisation, the need to tame 'globalisation' in the interest of sustainable regional employment and development is clearly addressed. Sustainable regional development is not on the global agenda. Independent, innovative, and enterprising owner-managers of small and medium-sized firms are identified as key players in regional development, as contrasted to corporate managers often mistaken in regional development policy for entrepreneurs but who are shown to have many characteristics of *bureaucratic, organisational man.*

Chapter Three outlines the basic importance for innovation and entrepreneurship at the grass-root level of noneconomic factors dismissed as 'friction' by mainstream, neoclassical economists. The innovative and enterprising potential of endogenous human capital is shown to be a function of modal personality, social structures, and cultural configurations in the region. This chapter heralds Chapters 4 and 5 which elaborate and document in detail the significance of noneconomic determinants of economic behaviour.

Chapter Four delves into factors identified as affecting individual socio-economic actors' disposition to innovative and enterprising activity; factors inherent to the individual's personality, to his or her socialisation and conferred values and attitudes, and to his or her specific social role and identity.

Chapter Five elaborates complex implications of the individual's embeddedness in socio-cultural contexts for hampering, encouraging, or triggering the actual use of his or her own innovative and enterprising potential and for outlooks for success in launching a new venture or maintaining an enterprise in less-than-optimal, if not hostile, oligarchical environments.

Chapter Six summarises the findings and draws some conclusions on basic reorientations in regional development policies, in particular with respect to the recognition of SME-entrepreneurs as the prime target group and to the need to stimulate regional incubators for independent creativity, in the interest of sustaining and augmenting grass-root fertility of endogenous human capital and activating it for innovation and new ventures. Among policy implications is the need to channel public sector resources away from incentive-granting commercial and industrial recruitment programmes aiming at enticing investments of exogeneous capital in the region toward stimulating endogenous potentials by meeting both material and immaterial needs of SMEs with regional bases of production and, in particular, of potential initiators of new ventures. These needs include open public access to modern transportation, and communication infrastructure linking them to global flows in information, services and component-parts and a receptive, 'incubating' socio-economic environment. Incentive granting programmes were shown to increase, rather than decrease regulatory and oligarchical environments to the comparative disadvantage of small enterprises and 'unconnected' outsiders, notably of potential initiators of new ventures.

The term 'regional development' implies an intention to change human behaviour and socio-economic contexts; and sometimes in the aftermath the cultural and physical environment as well. Certainly, experience shows that not all change affects our living conditions in a positive way. On the other hand, the term 'sustainable development' does imply that the envisioned regional circumstances will be - or, at least, are intended to be desirable, generally beneficial and, thus, worth sustaining. Hence, 'change', 'innovation', and 'development' will be given a positive connotation in this contribution on bottom-up strategies of regional development. Development is not conceived here as the annihilation or substitution of the genius loci in a deus-ex-machina approach, but rather as involving it in its own metamorphosis. Taking into account potentialities of regional socio-structural, economic, and cultural specificities at the grass-root level in their interaction with an exceedingly dynamic global framework may be just what is needed to provide local populations with the possibility to activate their particular social and cultural aptitudes in development processes. The genius loci, if viable, must come into play in socio-economic dynamics and evolve in the process. Our goal can hardly be to create regional 'open-air museums'. As can easily be observed in the world about us, the alternative to actively participating in development processes is to stagnate until being 'leap-frogged'.

Notes

[1]   'Sustainable rural development' is a term frequently used in work by and for European bureaucracy. In referring to sustainable regional development, we use the term in the wider sense, that is that the specific manifestation of regional development is both intended and long-lasting.

[2]   Smilor and Feeser 1991.

1   Goals and Policies for Regional Development:
A Critical View from the Grass-Roots

The most convincing and longest-living lies are half-truths. One such half-truth is that sustainable regional development depends on attraction of exogenous capital into the region. A very common community development policy for inciting capital investments ('commercial and industrial recruiting') is granting financial incentives (tax rebates, subsidies) to draw investors into target areas.[1]   Of course, no one would dispute that exogenous capital investments can and often do initially promote economic activity within target regions. With respect to sustainability, however, merits of induced investments from external sources are more than debatable. In many cases, induced investments prove to be, in the long run, outright disadvantageous. Quite frequently, it can be observed that internationally mobile capital interests cash in on incentives, bring in outside key personnel and create subsidiaries or subholdings. When regional markets cease to be of strategic interest or when more lucrative opportunities arise outside the region, they may subsequently transfer their assets to other subsidiaries, sell the most valuable assets and ultimately relocate their subsidiary, or simply declare insolvency of the regional holding and leave local bank loans and factories/other facilities behind.[2] Hence, in the process regions may very well lose many of their original assets or have them substantially depreciated. Consequences of a globalising economy for grass-root entrepreneurship are still not clear but many relevant effects of globalisation are discernible and will be outlined in this contribution.

Encouraging 'Spontaneous Combustion' of Entrepreneurship

Regional development policies in many developed countries aiming at attracting 'new' or established firms to disadvantaged areas have become less and less successful in recent years. One reason is that a very important push factor in (e.g., urban to rural) job migration, a shortage of labour at existing production sites, disappears in times of economic recession and unemployment[3] and is slow to reappear. In addition, many jobs in industrial production requiring low- or intermediate-level skills have been eliminated by improved production technologies or have been transferred to

1

countries with low wage levels. Transportation and communication technologies enable firms to flexibly relocate their component functions, to reorder entire production technologies, and to take advantage of world economies of scale in research and development, in raw material sourcing, and in production costs.[4] At the same time, economic output in goods and services has shifted in favour of services but even the distinction between goods and services has become blurred, almost meaningless.[5] Hence, regional policies are increasingly focussing on harnessing endogenous potentials of problem areas, specifically on stimulating local entrepreneurship, innovation, and technology.

Keeble and Wever[6] distinguish between two types of 'new' firms:
- firms new to the region but which are only subsidiaries/branch plants of already existing enterprises. (In the past, regional policy makers have been more prone to recruiting existing firms than to stimulating new firm formation.) and
- firms which did not exist before and which are set up by independent business(wo)men ('new ventures').

This distinction is important for regional development policy although, as we shall see in the following, these groups are by no means mutually exclusive.

Numerous agents are responsible to some extent for initiating and maintaining regional economic processes. On one side, there are politicians and public administrators aiming at providing infrastructure and aid. These socioeconomic actors have frequently been called 'social or community entrepreneurs' linking public and private sectors in the interest of community development.[7] As we shall see in the following, they are far from being 'entrepreneurs' and may well belong to the clique of 'good old boys' dominating the community. On the other, there are those directly involved in enterprise: corporate managers, owner/managers of small- and medium-sized enterprises (SMEs), and various self-employed professional groups.

For decades, there have been many attempts to define and describe clearly and succinctly the *innovative entrepreneur* as a member of the business community (private sector). Attention was given to allegedly specific personality traits, socio-economic characteristics, and roles of 'entrepreneurs' distinguishing them from the vast majority of the population. However, a clearcut delineation of the 'innovative entrepreneur' was not achieved because it was ascertained that even those traits which were shown to be important for entrepreneurship were not unique to any single group.[8] Data and intuition indicate that *indigenous* entrepreneurs are those most firmly rooted in their regions and are those least prone to relocation outside of the region. In fact, most entrepreneurs set up their own businesses in the general area where they live.[9] Cooper and Dunkelberg[10] note that about three-quarters of entrepreneurs do not move from their places of residence when starting their own firms. This suggests that nascent regional entrepreneurship is substantially dependent on the regional pool of 'human capital'[11] resources. Studies of 'regional success stories' invariably highlight the importance of entrepreneurial traditions, innovative spirit, energy, dynamism and qualifications of the local population.[12]

Another important factor in endogenous potential are existing regional economic structures, in particular a varied sectoral mix.[13] Regional economies with a large proportion of small- and medium-sized enterprises (SMEs) have lower entry barriers for new firms.[14] We shall not indulge here in attempts to delineate 'small', 'medium', or 'large'-size enterprises based on numbers of employees or size of capital stocks - which would vary, of course, according to the branch of the economy and the current state of technology. In referring to SMEs, we conceive them as having two essential defining characteristics outlined by Sweeney, namely

- they have not more than three, usually only one person making all major decisions, and
- the decisionmaker(s) is/are the major information processor(s) dominating or largely covering all external contacts.[15]

Empirical research has verified that most firm founders - and in particular the successful ones - come from other SMEs where they received better training for entrepreneurship than they could have acquired from more hierarchical, occupationally segmented establishments.[16] Small and medium-sized firms and those with regional markets are also less likely to relocate because they have much higher transaction costs of relocating operations than larger, multilocational companies. SMEs play a significant role in innovation[17] and smaller firms prove to be particularly flexible due to less specialisation and more opportunity for face-to-face contacts of economic agents.[18] They are reputed to be more able to adjust to rising demands within and beyond regional markets for more varied, sophisticated, and customised goods which are of higher quality and produced in short series.[19] According to Reich,[20] the potential for high profits and long-term survival in a globalising economy is shifting from firms concentrating on mass production ('high volume') to those with variable, high quality, customised production ('high value'). The value of such manufactured goods thus has a very large service component. Quoting Quah:[21] 'GDP will become increasingly weightless as economies grow progressively knowledge- and information-based. A succinct term for this is "dematerialization"'.

Quoting Van Praag and Van Ophem:[22] 'Governments are by and large of the opinion that new firm formation is necessary for a healthy economy and that the 'natural' entrepreneurship supply is insufficient. ...This calls for insight in the individual decision process. ...to identify ...determinants of both opportunity and willingness to become self-employed'. Self-employment is, according to these authors, closely related to but not synonymous with entrepreneurship. But it is not simply a matter of promoting self-employment as a means of alleviating the market for dependent employment by an equal number of individuals with independent gainful activity. For some time, creation of small and intermediate-sized firms (SMEs) had been seen as a major potential for job creation in economically advanced countries.[23] Although recent analysis indicates that SMEs have actually made only a relatively small contribution to employment, Keeble and Wever[24] predicted in the 1980s more significant long-term effects, noting that decline in employment had been greatest in larger enterprises while employment was at least increasing somewhat in SMEs. Such conclusions were more or less the result of particularities in classification of phenomena for data analysis. As Reich[25] emphasises, conclusions that large businesses are ceding to small businesses on the labour market or that manufacturing is being replaced by services ignores existing weblike relationships and problems in statistics. Hence, the question remains to be addressed again below whether an increase in the number of small enterprises not be simply a direct result of changes in organisational policies of large enterprises. Different forms of corporate governance create a major problem in identifying SMEs which really contribute to employment. This is due, for instance, to franchising policies of some major enterprises and to their externalisation ('outsourcing') of highly specialised or of repetitive and peripheral functions to subcontracting 'firms' (often consisting of former employees) with which they sustain stable relationships, frequently engendering the 'lion's share' of all business done by the up- and downstream firms. Although such subcontracting firms are not organised hierarchically within one corporate unit, they are in effect operationally part of what Granovetter terms a 'quasifirma'.[26] Reich[27] states that there is no 'inside' or 'outside' the corporation in such weblike business relationships, but only different distances from their strategic centres. He illustrates shifting centres of power of corporate and semi-corporate conglomerates by specifying four examples of corporate strategies:[28]

- Creating *independent profit centres* which exemplify 'lean management' concepts for eliminating middle-level management by delegating a certain degree of authority for production and service development, sales, etc., to groups of involved specialists whose work resources and personal compensation depend to some extent on their unit's profits.
- Establishing *spin-off* partnerships in which corporate headquarters buy up good ideas/technologies from external organisations or form strategic partnerships, then producing, distributing, and marketing the products or services under their known trademarks.
- *Licensing/franchising*, i.e., contracting with independent businessmen authorised to use the corporation's brand names and trademarks in return for adhering to special formulas and quality standards of the corporation, which also provides bulk services, such as advertising.
- Pure *brokering*, i.e., a loose form of collaboration with independent businesses for specified goals in research and development, resource-sharing production, distribution, or marketing.

It is precisely the explicit objective of juridical cleverness in organising business entities to escape typologies amenable to regulation and taxation. On the other hand, survival strategies and access to new technologies and markets may necessitate permanent or temporary strategic alliances with other firms or larger corporations. Another source of possible confusion for SME typologies are SMEs founded as a (joint) 'side-venture' of one or more entrepreneurs or consortia. Pluriactivity/multiple job holding as a practice and strategy in gainful activity has long been a subject of economic analysis and the advice of financial consultants to 'diversify' assets is well known. Entrepreneurs, too, have learned that diversification of their holdings, possibly with all or many of their enterprises within a general area of operations ('quasifirm'), renders not only more security of assets but gives them considerably more leverage with respect to regulations and taxation.

These aspects reveal the importance of distinguishing with Stöhr between

- *regional* development *policies*, which are defined as administrative measures taken 'from above', and
- *regional action*, which refers to mobilisation of development potential 'from below'.[29]
- Administration of regional development policy 'from above' has intrinsic macroeconomic orientation, such as achieving national and regional economic and social balance by taking general measures to reduce inter-regional disparities, and can only anticipate regional

exigencies collectively.   'Bottom-up strategies' have as common elements, according to Malecki and Tödtling[30] centrality of the role of entrepreneurship and of firm formation, centrality of technological innovation to enhance firm competivity, and importance of local and regional networks for entrepreneurship, new ventures and innovation.

Firm birth rates do vary significantly from country to country and from region to region.[31] Why do some regions or countries develop more rapidly than their counterparts?   Hagen[32] underlines the importance of technological creativity for productivity and economic growth. He sees economic and political circumstances in society as being closely related, non-economic aspects of human behaviour determining the state of technology and the rate of economic growth. And further:

> Economic growth has been led not by individuals distributed at random throughout the society but disproportionately by individuals from some distinctive social group.[33]

As we shall see in later chapters, some socio-economic and cultural contexts offer particularly fertile ground for 'entrepreneurial discovery'[34] and 'technological creativity'[35] and have been termed innovative environments ('milieux innovateurs'),[36] 'learning regions', and regions with 'local knowledge'.[37] Innovation is a driving force in creation of new ventures and in expansion of existing enterprises.[38] Innovations in today's world may well occur by 'technological leapfrogging' which does not fit well into models of dynamic innovation diffusion (e.g., based on the theory of epidemics) or of diffusion retardation (e.g., by already heavy investment in outdated technology - 'vintage capital').[39] Furthermore, firm foundation involves an inherent decision as to location, imposing geographical constraints on the organisation.[40] Examination and regional differentiation of the historical development of organisations by economists, sociologists, historians, anthropologists, etc., have revealed the strong impact of social and cultural forces on birth and transformation of organisations on all levels of society.[41] Innovations have failed when introduced to societies with nonsupportive cultural and institutional traditions. Hence, the importance of situational contexts for the study of innovation and dispositions to entrepreneurship.   These contexts may reflect differing prevalent values, attitudes, motivations, and images between populations, with different effects on modal personalities. These are topics of social-psychological investigation. Situational contexts may also manifest cultural differences, such as in knowledge, beliefs, and habits - a subject for social-anthropological study. But institutionalised and informal social structures and relationships between socio-economic actors of interest to sociologists are also important factors in innovation and enterprise.[42] 'Informal' structures and activities provide a frequent 'testing ground' for mobilising resources in both stable economic systems and in the adversity of transition economies.[43] All these factors may be considered to be significant non-economic determinants of microeconomic (grassroot) behaviour with meso- and macroeconomic implications.

Personal Factors in Self-Employment and Enterprising

As compared, for instance, to the situation in the USA, in highly developed industrial and service economies in Western Europe amenities of dependent work have indeed been quite considerable, even if now somewhat in decline. For instance: strict limits in personal responsibility and liability as well as in the number of work hours, relatively strong job security, personnel .representation, paid sick leave and work dispensation for numerous occasions, paid holidays and vacation, various fringe benefits for health and unemployment insurance and for pension and capital formation schemes, large Christmas gratuities, employee discounts, access to and use of the firm's recreational and vacation facilities, etc. In contrast, entrepreneurial self-initiative is immediately confronted with a host of regulations and liabilities in addition to the financial risks incurred. Has a socio-economic environment which would appear to reward dependency more than self-initiative in gainful activity gradually decreased the 'natural supply' of potential entrepreneurs in the population? Inversely, we may ask whether there is a hidden pool of entrepreneurs and how can we identify and activate them in regional development processes?

According to Meager[44] there has been renewed interest in self-employment in the past two decades after a long-term historical decline in most developed economies. Among the explanations, he cites:

- changing opportunities for dependent employment,
- sectoral change (decline of agriculture and manufacturing in favour of the service sector),
- changing aspirations (independence, urge towards creativity and self-fulfilment),
- trends in enterprises towards sub-contracting, i.e., towards externalising branches of their activities, and
- other factors, including government and business policies (e.g., early retirement schemes).

However, increase in 'self-employment' is certainly not totally synonymous with growing 'entrepreneurship' (and the creation of new jobs), especially in view of the fact that the majority of explanations offered are not related to a 'pull' towards entrepreneurship, but rather to a 'push' from the labour market. Bögenhold and Staber[45] warn against interpreting increasing numbers of small businesses and self-employed as a sign of macroeconomic vitality:

> Not all small economic units are innovative and flexible, many self-employed workers operate in marginal areas of economic activity, they generally suffer from high business failure rates and job instability and few of the newly founded small firms contribute significantly to job growth. The recent rise in self-employment may be a symptom of labour market deficiencies

Based on the recent rise in self-employment in the United Kingdom, MacDonald[46] concludes that much of self-employment is not a sign of a shift from 'dependency culture' to 'enterprise culture' but rather 'survival self-employment' as part of a growing culture of informal and risky work. Many of the newly self-employed had little previous business experience and were from social groups previously underrepresented in the small business population, like women and young people. Accordingly, there were few successes. In addition, in spite of proclamations of policy makers to the contrary, individuals inclining towards entrepreneurship find themselves in a more and more hostile institutional environment, even if the hostility is not intentional. As a result, in recent times individuals genuinely and naturally inclining towards entrepreneurship would appear in modern industrial society to still be an endangered species.

Fromm[47] suggested that modern industrial man is characterised by a desire to 'escape from freedom', i.e., to let others make his decisions for him and take responsibility. Riesman[48] outlined an evolution of the relationship between individual character and society as a process leading

- from *tradition-directedness* of behaviour based on strongly enforced ascribed status, values and roles of the individual
- to *inner-directedness* (autonomy) in line with trends to individualisation, personal mobility, technological progress in transportation and communication, etc.,
- and ultimately in highly developed, materially saturated western societies:
- to *other-directedness*, i.e., reliance for guidance on peers and on mass media.[49]

However, as Lerner[50] and Hagen[51] have pointed out in their typological concepts of social change, at any point in time in history there is a great plurality of forms and traits of societies in transition from 'traditional culture', held to be custom bound, hierarchical, ascriptive and unproductive, to 'modern/post-modern culture', said to be characterised by strong geographical and social mobility, significance of the industry and service sectors, high rates of literacy, urban ways of life, openmindedness, powerful inclination toward rationality, and strong social participation of the population. This then would be the basis of intercultural differences in the given prevalence of certain modal personality types and value systems and hence in a specific culturally determined propensity of individuals toward or against technological and socio-economic change.

General Concepts in Regional Economics and Development Policy

As a rule, macroeconomic goals of economic stability and vitality pertain to and depend on the sum of circumstances on the mesoeconomic level, i.e., in regional economies. Development measures are thus applied to bolster and promote market dynamics in stagnating or lagging regional economies. For instance, improvement or renewal of infrastructure is intended to have secondary effects promoting regional production and employment. Measures taken for regional development are manifold: There are direct subsidies, such as state and supranational subsidisation of investments (e.g., investment subsidies, credit subsidies, reduced interest rates, risk securities, research and development aid) and local and regional support (e.g., reduced siting (land) costs, reduced rents and leases, lower local taxes and fees, respites in payment of taxes and fees). On the other hand, there are many indirect support measures. These include improvements of sites and infrastructure (for transportation, communication, public services, general environment, commercial and industrial centres and expositions sites, technology centres, educational, recreational, and living facilities), measures to augment the professional qualifications of the local population (local competitions in science and technology, linkages between business and local educational institutions, establishment of technology transfer agencies), promotion of professional associations, provision of consulting and marketing services.

Neoclassical growth models have focussed attention on capital formation and, particularly in regional applications, on capital mobility, i.e., bringing more capital into the region.[52] To attract capital investments, regional economists propose measures for maintaining and utilising regional production cost advantages and local resources, and for improving access to markets. Innovation oriented regional development policy of national and supranational authorities thus aims at influencing siting decision making by potential investors and entrepreneurs in favour of locations in target regions.[53] For quite some time it has been obvious, however, that losses in industrial employment are no longer an indicator of sinking levels of industrial productivity, output, competitive capability or of decreasing market shares and profit margins.[54] It is also obvious that traditional location theory (postulating rational and deliberate weighing of costs and benefits of alternative sites) is totally inapplicable to the newborn firm phenomenon. This has incited the elaboration of theories on firm incubation and propitious characteristics of the environment for both existing dynamic firms and new ventures, including attitudes and qualifications of the population as well as structural characteristics of the regional economy and society.[55]

On the other hand, *endogenous growth theory*, stressing reliance on regional potentials for sustainable economic growth, has as a central postulate that in order to incite economic growth marginal productivity of regionally accumulated capital must not decrease.[56] Romer, one of the pioneers of 'endogenous growth theory',[57] presented a model of long run (sustainable) growth in which 'knowledge' is assumed to be a factor input with *increasing* marginal productivity.[58] Knowledge or know-how was thus recognised as a basic form of human capital which is very important for endogenous regional growth. The chosen indicators of this endogenous regional capital were levels of schooling and training.[59] As we shall see in the following, this is only a partial view of grassroot potential. Nevertheless, in models of endogenous growth economic actors are expressly treated as more than simply as rational resource allocators and human capital is recognised as potential for increasing marginal utility.

Sustainable regional development has now come to be widely perceived as being largely dependent on economic agents operating 'in situ' and development prospects as being a function of their attitudes and behaviour. Two major conditioning factors in situational contexts of human capital specified by Molle are:

- access to ideas and flows of information and
- innate receptivity of the population to new ideas.[60] Here we begin (but only begin) to really scratch at the grassroots.

The key to economic growth in contemporary, highly competitive, and technology oriented society is innovation. By consequence, in analysing situational contexts of human resources, comparative locational advantages or handicaps with respect to innovation must be a central point of interest. Regional economic theory has a number of concepts for dealing with this phenomenon. For instance, in placing innovation life cycles in their geographic context, Molle[61] elaborates regional 'filtering-down theory' by enumerating the following stages of innovation:

- growth phase: innovation of products and processes at its peak necessitates sites providing optimal information and communication facilities,
- ripening phase: gradual standardisation of new products and processes with diminishing needs for information and communication but increased needs for market access, and
- stagnation phase: advanced standardisation tending towards locations with lowest production costs.

Understandably, attention generally focusses on conducive environmental traits for the least known, yet most important initial (growth) phase which triggers the whole process. Since innovation processes are a function of the specific technologies involved, Orsenigo[62] stresses the significance of *'technological regimes'* in locational analysis. Noting general cumulative benefits of innovation (i.e., today's innovations are the starting point for tomorrow's innovations), Orsenigo points out that *new knowledge may be more or less pervasive* in the sense that it be applicable to either a variety of products and processes or just to a specific technological set.

Judging by the variable frequency of technological innovations and differing rapidity in their diffusion, some regions appear to serve especially well as innovation incubators or as particularly fertile receptors. Taking a territory-based and grassroots view of innovation processes, the GREMI-research group,[63] in trying to identify *'des milieux innovateurs', 'l'esprit innovateur local', 'des synergies locales',* focusses on creative activity in given social environments which meets local challenges and makes use of local experiences.[64] The group has determined that while innovation breaks with the past, it also must rely on cumulated elements of the past (e.g., knowledge base in specific technologies) to facilitate its insertion into the field, to reduce resistance, and to increase its diffusion potential. Among indispensable elements of the past are the specific quality and structures of human relations in the region. Territorial dynamics are obviously a function of interactions and networks between economic agents. With reference to the *'transaction cost'* concept introduced by R. Coase and further elaborated by Williamson and others, the GREMI-group underlines the *significance of proximity and of 'informal'* (i.e., non-contractual) *relationships between economic actors in minimising transaction costs.* Examples are given of strongly differing territorial dynamics due to regionally specific atmospheres in human relations. This provides us with considerably better insights into how the grass grows. Our efforts in this contribution will be directed at deepening insights into the structures and functioning of grassroots along these lines.

Reynolds, Storey, and Westhead[65] differentiated current policies and programmes with the aim of influencing the entrepreneurial process according to whether they aim at encouraging conception of new firms, at facilitating their gestation by improving public infrastructure or by offering general programmes to inform and train potential entrepreneurs, or at facilitating the growth and survival of enterprises by providing resources or specific counselling and communication networks. The authors' key finding was that a high proportion of the regional variation in firm birth rates in the European Union can be explained or predicted by appreciating regional characteristics. National, regional, and local milieux are shown to have a profound influence on new firm foundation and economic growth.[66] Reynolds, Storey, and Westhead proceeded to investigate causes of regional differences in firm birth rates on the basis of regionalised macroeconomic data, e.g., by using indicators of 'demand growth', urbanisation/agglomeration, unemployment, personal household wealth, firm structural characteristics, electoral behaviour and government measures. Their analyses led to a 'suspicion' that different countries in Europe have significantly differing degrees and patterns of encouragement of entrepreneurship from the social and institutional environment.[67]

An endless number of political, administrative, and scientific texts, meetings, and projects have ascertained the hardly surprising fact that economic activity is not equally distributed in space. Also hardly novel is the insight that regions are unequally endowed with potentials and that rural space is not monolithic. Obviously, modern industrialisation had centered on urban agglomerations and/or transportation nodes. Rural areas were in general economically residual areas dominated by weakening primary sectors such as agriculture, forestry and mining. This was the common feature of most rural areas, not their potentials. Certain locational potentials, notably natural beauty and unspoiled resources, have become more relevant with developments in communication and transportation technologies. In countries with relatively dense infrastructural networks, such as Germany and the Netherlands, even decentralisation of some industry from urban agglomerations (most often to rural centres) occured but it was especially the transition to a service-based economy which aroused hope for rural development. However, in view of effects of technological rationalisation of work processes and of job expatriation from countries with high wage levels, expectations of full substitution of redundant industrial jobs by the service sector have been unrealistic. But the shift to a service-based economy has certainly augmented the significance of human capital for rural development.

In spite of the sharp decrease in importance of the agricultural sector for the Gross National Product and for labour markets in most countries, even in rural areas, most national and supranational (e.g., EU) policies and programmes for rural development and research still are heavily dominated by the agricultural sector.[68] This manifests itself in the prevalence of institutions and agents of the agricultural sector in formulation, implementation, and evaluation of policies, programmes, and research projects for rural development. In practice, to solve logistical problems and to save transaction costs, it is often advantageous to make use of existing institutional networks. Nevertheless, we shall see in the following that institutions represent specific normative systems and power relations of given interest groups. In their endeavour to maintain regional balances they are a potent stabilising factor for existing, possibly obsolete structures. Keeping in mind the extremely small and decreasing fraction of gainfully active populations in all industrialised countries still engaged to a significant extent in farming, even in rural areas, most of the (rural) population will not have the impression of being a target group of agricultural rural development programmes, i.e., of policy by and for the agricultural sector and, in fact, mostly they are not, neither explicitly nor implicitly. Despite major idiosyncrasies in their strategies for income and allocation of resources, for social security and in policy dependency, risk-taking, etc.,[69] farmers are frequently equated with small firm entrepreneurs. The resultant scope of development concepts in rural research and development policy is accordingly rather limited. For instance, many programmes seek alternative sources of work income only for members of farm families, concentrate on agro-tourism, overly highlight alternative uses of farm buildings and machinery, often even excluding the nonfarming population from programme benefits, e.g., eligibility solely of the farming population for financial aid in creating/improving/maintaining guest accommodation and tourist infrastructure. Thus, institutional factors have a decisive impact on both the scope of policy and target groups for rural development.

Merlo and Manente[70] also noted that, in early phases of rural development policy, rural development was considered to be more or less synonymous with development of the agricultural sector. With realisation of decreasing economic importance of the farm sector even in rural areas, an interest arose in 'integrated' rural development, aiming at more comprehensive, multisectoral regional development. (Institutional structures still remained, however.) They differentiated development policy according to whether it used an intersectoral approach (i.e., based on input/output analysis), a regional economic approach (analysis of opportunities and constraints due to geographic, demographic or economic features of locations), or a socio-economic approach (analysis of perception and behaviour of social groups/actors). The socio-economic approach is taken here to examine the grassroot fabric (human capital) with respect to its social and economic cohesiveness and its adaptive capacities within a high-tech environment and a new international order.[71] The significance of this research topic is highlighted by hypotheses put forth by Camagni[72] to the effect that least favoured EU regions will benefit less from the Single European Market and will suffer more from harmonisation of regulations and lifestyles than their better-off counterparts and that they will rapidly lose their initial cost advantage in unit labour. Camagni concludes that regional development is in need of new principles and philosophies rather than just new tools.

Current Role of the State

In modern industrial and 'post-modern'/'second modern' service-based society, no matter what the current national economic philosophy may be, the population invariably looks to the state for economic security and its demands in this regard remain at a high level, even if fluctuating somewhat in the course of time and varying in scope.[73] In basic terms, economic security implies that the socio-economic system assure gainful opportunities for all members of the working population, that it provide for needs of the economically non-active population, and that such economic activities guarantee a level of living in accordance with present day socially accepted standards of living.

Kent[74] enumerates the following generally accepted economic roles/ goals of government:
- maintenance of internal stability and enforcement of legal rights
- defence of the nation from external aggression
- provision of public goods (e.g., police, fire departments, parks, roads, flood control)
- expansion of consumption of goods with high external benefits (education, public health)

- prevention of external costs by taxation, prohibition, or regulation (e.g. pollution, depletion of natural resources, establishment of standards)
- regulation of natural monopoly
- counter-cyclical fiscal and monetary policies and
- redistribution of income.

What is subject to dispute, however, is how the state can best attain these goals, in particular, to what extent the state should intervene in economic mechanisms. For many decades now, western economic stewardship has followed paths indicated by John Maynard Keynes in order to stabilise national income and employment in the face of strongly fluctuating economic cycles, for instance by regulating the quantity of currency in circulation and by the use of public expenditures to bolster the national economy in periods of stagnation/recession. But countless other measures have also been taken by governments in conjunction with 'social partners', i.e., institutions representing employers and the working population ('collective bargaining'), to guarantee economic and social stability and welfare.

In general, benefits to society and the majority of individuals derived from laws and regulations on taxation, labour conditions, etc., are indisputable. However, as we indicated above, for some time it has been apparent that an institutional environment has gradually evolved which hampers individual spontaneity, initiative, and flexibility.[75] In a recent article,[76] Helmut Schmidt, the former German chancellor, claimed outright that an overregulatory environment was suffocating the spirit of entrepreneurship. Taking as a concrete example the situation which would be encountered by a talented and willing to be entrepreneur in the field of electrotechnology in Germany, Schmidt illustrated his point by presenting a list of German laws, regulations, ordinances, etc. to which the entrepreneur's actions must conform and for which he or she would be held accountable. Excluding introductions, tables of content, charts, graphs, commentaries, the list included the following texts:

- laws for employee protection (*Arbeitsschutzrecht* - 314 pages)
- employment promotion act (*Arbeitsförderungs-Gesetz/Verordnungen* - 359 pages)
- income taxation regulations (*Einkommenssteuerrecht* - 510 pages)
- social legislation, e.g., for child benefits and allowances, accident insurance, education and training, safety and accident prevention, housing allowances, health insurance, pension, insurance, social compensation, rehabilitation of disabled persons, social assistance, long-term care insurance, workers' say in operations ('codetermination'), maternal benefits (*Sozialgesetzbuch* - 1533 pages)
- environmental regulations (850 pages)

- tax laws and regulations of townships (637 pages) and,
- building codes (1731 pages).

rendering a grand total of approximately 6000 pages of legal paragraphs which, as is the very nature of legal texts, are subject to and are the subject of myriad interpretations and innumerable judicial decisions in litigation, documented in sheer unending texts. Even if the potential entrepreneur had the means for financing a team of specialised legal experts, costs in time and the lack of predictability of outcome in cases of litigation make the potential entrepreneur's electro-technical competency and knowledge of the market seem, in comparison, rather insignificant in view of entrepreneurial success. It is a well known fact that large amounts of entrepreneurial resources are frequently and successfully 'invested' in finding loopholes in such legislation. Success in finding loopholes not only undermines the very purpose of the legislation, it also puts large enterprises with legal staffs at a competitive advantage over SMEs. Thus, *overregulation distorts competition to the detriment of SMEs and of new enterprises without specific juridical know-how and information networks.*[77] It does not, by any means, effectively channel resources into technical innovation or technologically competent, market-oriented entrepreneurship.

In dependency culture, the entrepreneur is frequently considered by many to be a potential exploiter of labour and a profiteer, rather than as a potential job creator. Similarly, investment in rental housing - another formerly widespread form of private investment, e.g., for personal financial security in retirement, which benefited primarily the locality and region - is in most cases not very lucrative and very work intensive, to say the least. 'Landlords' are also often equated with exploiters and profiteers. It would thus appear to be socially more acceptable and comfortable in many developed economies to place personal investments in the global financial market rather than to be a 'capitalist' entrepreneur or landlord in the region. Dependency culture has accentuated this move to 'clean'[78] capital markets yielding high interest: Investing in one's own enterprise, especially in one creating employment for others in the region, may entail a several year long struggle just to 'break even'.

Supiot[79] observed that the 'welfare state' had led to a steady growth of persons working for the public sector, whilst state control over employment relationships in the private sphere had also increased. Counterreactive efforts at privatisation were undertaken in Western Europe and North America in the 1980s and 1990s to reduce the public sphere. Close scrutiny shows that these efforts have, at their exceptional best, been able to diminish the number of persons working directly for the state but the main result has merely been a restructuring of relations and a shift in functions between the state and private spheres but without effectively reducing the state's role. Much more important is the fact that possible declines in the number of public employees has been accompanied by a rapid expansion of the state's regulatory function.[80] Deregulation measures have proved to be ineffective; a strong regulatory environment persists.

The above comments address constraints put on entrepreneurial behaviour by a regulatory environment. But not all institutional measures are taken to constrain entrepreneurial behaviour; some obviously aim at creating opportunities. We have already mentioned diverse measures that promote initiative and investment in economic activity. State policy to encourage regional development includes measures to subsidise investments and loans, to grant low interest rates and securities, and to subsidise research and development activity. Measures taken by local communities and counties include reductions in costs of land and rent, reductions in local taxes and service fees, respite of loan payments and of tax dues, low cost provision of transportation, educational, technical and consulting infrastructure/services, etc. As a matter of fact, such instruments for regional development enjoy increasing popularity with public administrators as well as with benefit recipients. These measures appear to be only restrained by public administration's increasingly limited access to funds. It is, of course, much more pleasant to distribute candy than bitter pills and more enhancing for self-representation of politicians and administrators to bestow grants than to impose constraints and penalisation. Nevertheless, putting it in economic terms: there is no value added by (re-) distribution of resources. Putting aside the question of abuse of government funded subsidies,[81] we shall attempt to illustrate in following chapters that the administrative art of *(re-) distribution of resources is not propitious to creativity, technical competence, or innovation in the economic sector but access to these resources does make enormous demands on work resources and networking skills of potential innovators - which would be better used for innovating activity and probing and winning new markets.* To make things worse, it has become all too apparent that structures of income generation and of state mechanisms for income redistribution have, in the context of globalisation, become totally incompatible.[82] As we shall note in the upcoming chapter, fewer products and services have distinct 'nationalities'. What is produced and traded is ever more frequently international composites of materials and services, making taxation and regulation extremely difficult.[83]

Once again: although consensus exists that government regulation of economic processes by creating constraints and opportunities is - to an undetermined, but limited extent - beneficial to and necessary for economic stability and social welfare, resource inputs by governments and enterprises in conjunction with these measures are, in themselves, non-productive. They do not necessarily or primarily stimulate grassroot innovation and nascent entrepreneurship and, furthermore, are becoming increasingly ineffective for social welfare. It is our contention that not only overregulation, but much of financial incentive dispensing policies are to a great extent even counterproductive to creativity, innovativeness and entrepreneurship at the grassroot level.[84] Recent research has shown that, at best, financial incentives and tax abatements have proven to be only secondary motives in locational decision making of business interests. Their effects were found to be clearly less important than the presence of modern infrastructure and public services, some studies find that even quality of life factors are more likely to influence decisions.[85] Financial incentives are, of course, welcomed as fringe benefits but it appears that their major effect is symbolic (the local community can be expected to be very supportive). The handicap is particularly severe for existing SMEs and for recently founded firms with less developed personal and informational networks, which may be primarily oriented to regional markets and whose economic base is a single product or service or a closely related set. Nevertheless, ever increasing proportions of both state and entrepreneurial resources are absorbed by activities involving regulatory and redistributative mechanisms. Only multinational firms have succeeded in circumventing or dealing effectively with the regulatory environment of modern national economies.

The steady rise of multinational firms, rapid globalisation of financial, consumer, and labour markets, and evolving supranational bases of jurisdiction and litigation tend to further concentrate power in supra-regional and supranational oligarchies with extremely well developed telecommunication networks for quick access to information and resources. In global finance markets, decisions to withdraw capital/loans/assets from a regional holding may come faster than decisions to invest and these decisions may well have little or no relationship to the economic function or viability of individual local enterprises or subsidiaries in question and certainly not with social and economic welfare in the region but rather with 'power politics'.[86] Sustainability of regional development certainly plays no role whatsoever in such cosmopolitan decision making - unless, by chance, members of power oligarchies have themselves vested (personal or political) interests in the region. If this fact would seem to highlight advantages of close links between political decision makers and the world of big business and finance, the reverse side of the coin is that closely knit, supralocational oligarchical networks have the capacity to formulate and have 'cronies' in public administration implement incentive policies tailored to their own specific needs and to cash in on these benefits ahead of 'unconnected competitors'. This certainly must have stifling effects on budding regional endogenous human potential.

In the following chapter we shall take a closer look at the overriding economic framework of our time: highly accelerated globalisation of the economy, at increasing impotency of democratic institutions on the national and regional levels to deal with it, and at implications for regional endogenous potential for innovation and entrepreneurship. In subsequent chapters, we shall then focus our attention on human capital as the decisive component of endogenous regional potential[87] and describe basic mechanisms in human economic behaviour determining the existence of fertile environments for innovation and enterprise. This grassroot perspective will reveal that the incentive-based approach to regional development as a rule not only favours well connected, supralocational 'outsiders', but also tends to reinforce existing but stagnating or obsolete structures, putting the unconnected but truly entrepreneurial or innovative personality in the region in an unfavourable competitive position.

Notes

[1] Cf. Malecki 1998; Sweeney 1987; Gabe and Kraybill 1998; Loy and Loveridge 1998; Zimet and Lachter 1998; and Rainey 1998.

[2] Cf. Michalos 1997, p. 34 and Markusen 1985, p. 283. Loy and Loveridge (1998) find 'industrial and commercial recruitment' with the help of direct incentives to be the most popular strategy for local development groups in the USA and that it is getting more and more expensive per job created while the duration of the jobs in question is uncertain. For instance, citing Ch. Mahtesian (Romancing the smokestack. Governing (8) 11/1994, 36-40), the authors point to a grant of 71 Mio. dollars to Volkswagen by the State of Pennsylvania in 1978 for the promise of 20 000 jobs which did not exist ten years later. According to the experience of Zimet and Lachter (1998), evaluation of public expenditures per job created in industrial recruitment is seldom possible since politicians and local administrations are prone to not keeping records.

[3] Cf. Keeble and Wever 1986, p. 1.

[4] Cf. Rodwin and Sazanami 1991, p. 4.

[5] Reich 1991, p. 85ff. and Rodwin and Szanami 1991, p. 4.

[6] Keeble and Wever 1986, p. 2f and 29.

[7] Cf. for instance Stöhr 1990; Malecki 1994, p. 131 and 1998;Spilling 1991, p. 34; and Cromie et al. 1993, p. 250.

[8] Malecki notes that the 'entrepreneur', as a key economic actor, is missing in the modern theory of the firm. Cf. Malecki 1994, p. 120.

[9] Keeble and Wever 1986, p. 7.

[10] Cooper and Dunkelberg 1987, p. 19.

[11] According to Lazonick (1991, p.177) it was Theodore Schultz who introduced this term in 1961.

[12] Cf. Rodwin and Sazanami 1991, p.26; Simon 1996, p.10; and Keeble and Wever 1986, p.18ff.

[13] Cf. Sweeney 1987, p. 103; and Malecki 1994.

[14] Sweeney 1987, p. 11ff and 16f.

[15] Sweeney 1987, p. 145f.

[16] Keeble and Wever 1986, p. 6.

[17] Keeble and Wever, 1986; Jarillo 1989, and Simon 1996, p 7f.,

[18] Cf. Storey in Curran et al. 1986, p. 90; Simon 1996, p. 6f; and in particular the comparison of (dis-) advantages of SMEs and large enterprises presented in Rothwell and Dodgson 1991, p.127.

[19] Cf. 'success stories' in Emilia Romagna/la Terza Italia in Brusco 1982 and Bagnasco 1977.

[20] Reich 1991, p. 82ff.

[21] Quah 1996, preface.

[22] Van Praag and Van Ophem 1995, p. 513. As we shall see, the authors' data analysis disclosed that there are far more individuals willing to become self-employed than there are opportunities. (p. 530)

[23] Gibb 1993; Sweeney 1987, p. 8ff; and Davidsson et al. 1993, p. 150. Bannock (in Curran, et al. 1984, p. 14) cites research disclosing that 66% of new jobs in the U.S. service sector were generated in the period 1969-1976 by small firms. According to Bannock, (p. 17): 'people who work in small firms are much more likely to set up a business of their own than employees in large firms' Cf. also Reynolds, Storey, and Westhead 1994, p. 444ff; and Binks and Jennings in Curran et al. 1986. Martin (1984) also cites a number of studies pointing to the vital role of small and innovative companies in the process of technical change and for generating economic growth and employment. They quite frequently assume the function of 'incubator organisations' due mainly to factors connected directly or indirectly to human capital (p. 259).

[24] Keeble and Wever 1986, p. 23ff.

[25] Reich 1991, p. 95.

[26] Granovetter 1985, p. 497ff.

[27] Reich 1991, p. 96.

[28] Reich 1991, p. 91ff. Rothwell and Dodgson (1991, p. 126) list many more modes of web-like business interactions between SMEs and large corporations.

[29] Cf. Stöhr in Aydalot 1986.

[30] Malecki and Tödtling 1995, p. 280.

[31] Sweeney 1987, p. 1, 47ff, and 198ff; Davidsson et al. 1993, p. 146.

[32] Hagen 1962, p. 49.

[33] Hagen 1962, p. 28.

[34] Kirzner 1997. Kirzner describes 'entrepreneurial discovery' as the driving force in a process overcoming participants' 'sheer ignorance' of supply and demand mechanisms of the market system. Some forms and intensity of market interaction of economic agents are considered more propitious to the enhancement of knowledge about the market system and its needs.

[35] Hagen 1962. The author describes 'technological creativity' as the discovery of new

knowledge or concepts and their incorporation into productive processes. This second step he considers to be 'innovation'.

36 Cf. e.g., Aydalot 1986 and Camagni 1991.
37 Cf. Simmie 1997; Bennett and McCoshan 1993; and Malecki 1998.
38 Sweeney 1987, p. 10.
39 Cf. Soete 1985; and Chari and Hopenhayn 1991.
40 Cf. Pennings in Kimberly, Miles et al. 1981, p. 135.
41 Cf. Aldrich 1979, p.21f; and Halstead and Deller 1997.
42 Cf. Hagen 1962, p. 84f.
43 Cf. Roberts and Jung 1995, pp. 53f and 75ff.
44 Meager 1992, p. 87.
45 Bögenhold and Staber 1991, p. 224.
46 MacDonald 1996. 'Unwilling entrepreneurs' form a large proportion of new venturers in both Western and formerly socialist economies. Cf. e.g., Tichonowa 1997 and Oloffson, C.; Petersson, G.; and Wahlbin, C., (1986), 'Opportunities and Obstacles: A Study of Start-ups and their Development'. in Ronstadt et al.
47 Fromm 1941.
48 Riesman 1950, also cited in Parsons and White in Lipset and Lowenthal 1961, p. 89ff.
49 Riesman's concept certainly offered fertile ground for Inglehart's concept of changing values and attitudes in the change from *traditional* to *modern*, to *post-modern* society.
50 Lerner 1958.
51 Hagen 1962.
52 Cf. Grossman and Helpman 1994, p. 25; Eisinger 1988, p. 12; and Amin and Thrift 1993.
53 Cf. Ciciotti 1987, p. 298f and Steinnes 1984.
54 Cf. Rodwin and Sazanami 1991, p. 29.
55 Cf. Keeble and Wever 1986, p. 28.
56 Cf. Paqué 1995, p. 239ff.
57 Cf. Pack 1994, p. 55 and Grossman and Helpman 1994, p. 35.
58 Romer 1986; see also Romer 1994.
59 Cf. Grossman and Helpman 1994, p.35.
60 Molle 1987, p. 110f.
61 Molle 1987, p. 112f.
62 Orsenigo 1995, p. 43.
63 Aydalot 1986; and Camagni e Cappelin in Camagni, Cappelin e Garofoli 1987, p. 131ff.
64 Cf. Courlet and Pecqueur 1991, p. 309. See also Maillat 1995. Malecki (1998) citing concepts of Geertz and Asheim focuses on the role of 'local knowledge' for creating 'learning regions'.
65 Reynolds, Storey, and Westhead 1994, p. 445.
66 Cf. also Camagni 1992, p. 364.
67 Reynolds, Storey, and Westhead 1994, p. 453.
68 Some literature revealing this state of affairs will be offered in the literature references. For instance, large amounts of funds are allocated to the so-called EU-LEADER programme proposing to encourage local initiatives in enterprising. Demands for more funds for this programme persist although serious conceptual foundations are lacking and common denominators in methods used in the many

participant countries are extremely small; indeed, the EU conducts 'surveys' among programme participants to determine just how funds are used. Cf. EU-LEADER (www) and Chanard (www). The EU-ORA programme (rural technological infrastructure) is much more efficient.

[69] See Cécora 1991b.

[70] Merlo and Manente 1994, p. 135.

[71] Cf. European Commission: The future of rural society. Communication 371, Supplement to Bulletin of the European Communities 4/1988.

[72] Camagni 1992.

[73] See e.g., the 'Beliefs in Government' series published by Oxford University Press in 1995.

[74] Kent in Kent, Sexton, and Vesper 1982, p. 248f.

[75] Cf. e.g., Sweeney 1987, p. 14f.

[76] Schmidt, H.: Der Paragraphenwust tötet den Unternehmergeist. Die ZEIT (15), April 4, 1997, p 3.

[77] According to Brusco, investigation has shown that in the economically most dynamic area of Italy ('la Terza Italia') the state plays a lesser role in both spending and taxation than in other regions. Cf. Brusco 1982, p. 183.

[78] 'Clean' investments might well be positioned, for instance, in the armaments or biochemical warfare industry directly or by way of mutual funds or other investment packages.

[79] Supiot 1996, p. 653ff.

[80] Cf. Schmidt 1997 and Supiot 1996, p. 653ff.

[81] Cf. Waits and Heffernon 1994.

[82] Cf. Radermacher 1997, p. 3. See also Paci's findings from interregional comparative analysis in the European Union that convergence in economic indicators is concurrent with continuing large disparities in economic welfare .

[83] Cf. Reich 1991, p. 112ff.

[84] Of course, there are those who would contradict this view. For instance, scientific representatives of the (richly subvention-endowed) agricultural/rural sector who go so far as to theorise on 'induced institutional innovation' brought about by wise 'political entrepreneurs'; see Ruttan and Hayami 1984.

[85] Cf. Rainey 1998; Halstead and Deller 1997;and Zimet and Lachter 1998, as well as the literature cited in these papers.

[86] Cf. Michalos 1997.

[87] According to Radermacher (1997, p. 9), a World Bank study assessed 60% of national resources of states to be 'human', the other 40% to be equally divided between raw materials and manufactured goods.

2 The Global Economy:
The Sea on Which National and Regional Economies Sail

The key element in the 'global economy' is not production of goods or provision of services but rather the financial market. For this reason, it would be appropriate to commence the discussion thereof by quoting S. Strange in her book on 'casino capitalism' in the mid 1980s:[1]

The Western financial system is rapidly coming to resemble nothing as much as a vast casino. Every day games are played in this casino that involve sums of money so large that they cannot be imagined. At night the games go on at the other side of the world. In the towering office blocks that dominate all the great cities of the world, rooms are full of chain-smoking young men all playing these games. Their eyes are fixed on computer screens flickering with changing prices. They play by international telephone or by tapping electronic machines. They are just like the gamblers in casinos watching the clicking spin of a silver ball on a roulette wheel and putting their chips on red or black, odd numbers or even ones.

As in a casino, the world of high finance today offers the players a choice of games. Instead of roulette, blackjack, or poker, there is dealing to be done - the foreign exchange market and all its variations; or in bonds, government securities or shares. In all these markets you may place bets on the future by dealing forward and by buying or selling options and all sorts of other recondite financial inventions. Some of the players - banks especially - play with very large stakes. There are also many quite small operators. There are tipsters, too, selling advice, and peddlers of systems to the gullible. And the croupiers in this global financial casino are the big bankers and brokers. They play, as it were, 'for the house'. It is they, in the long run, who make the best living.

These bankers and dealers seem to be a very different kind of men working in a very different kind of world from the world of finance and the typical bankers that older people remember. Bankers used to be thought of as staid and sober men, grave-faced and dressed in conservative black pinstripe suits, jealous of their reputation for caution and for the careful guardianship of their customers' money. Something rather radical and serious has happened to the international financial system to make it so much like a gambling hall.

What is certain is that it has affected everyone. For the great difference between an ordinary casino which you can go into or stay away from, and the global casino of high finance, is that in the latter all of us are involuntarily engaged in the day's play. A currency change can halve the value of a farmer's crop before he harvests it, or drive an exporter out of business. A rise in interest rates can fatally inflate the cost of holding stocks for the shop-keeper. A take-over dictated by financial considerations can rob the factory worker of his job. From school-leavers to pensioners, what goes on in the casino .. is apt to have sudden, unpredictable and unavoidable consequences for individual lives.

This cannot help but have grave consequences. For when sheer luck begins to take over and to determine more and more of what happens to people, and skill, effort, initiative, determination and hard work count for less and less, then inevitably faith and confidence in the social and political system quickly fade.

The new economic order to which national and regional economies are subject is characterised by international business and financial centres in which multinational firms with enigmatic structures of ownership, affiliations, and control strive to take advantage of scale-economies and compete for ever-larger shares of global markets. These transnational enterprises are served, on one hand, by world-wide credit markets making use of a whole new set of financial agencies, methods, and instruments, and on the other, by a multitude of specialised services enabling them to regulate their own interactions and litigate their disputes in private self-governance on the supranational level and to penetrate and conquer national markets. On the national level, many of these specialised service agencies to transnational firms and financial institutions now assure as specialised subcontractors functions formerly assigned to firm employees or provide consulting, e.g., on property rights, on anti-trust legislation, on formulation of contracts, on laws for consumer-, labour- and social-welfare, and for protection of the environment, on accounting, on personnel management and training, on market analysis and forecasting, on production and communication technologies, on tax and subsidy policy, on advertising and public relations, on monetary policy of central banks and on political lobbying, etc. On the international level specialised services implement commercial arbitration between firms, mergers, and acquisitions. They also provide information on credit and interest rates, currency exchange rates, securitisation, values of stocks and bonds, and serve as brokers. When large enterprises externalise specialised functions, with-draw from peripheral activities, or resort to franchising policies they contribute to an increase in the number of SMEs without relinquishing most of their control. Very successful new SMEs are, as a rule, ultimately acquired by larger consortia or merge with other entities, augmenting the concentration of economic power.[2] And even the largest firms are prone to making 'strategic alliances' with potential competitors to enlargen the infrastructure, material resources, and pools of technology and human capital at their disposal. The often ephemeral web-like structures of transnational 'quasifirms' are, in effect, practically fathomless for juris-diction and taxation. However, they function very effectively as channels for rapid concentration of economic power.

International commercial and financial interest groups frequently make use of low-wage labour in underdeveloped countries and of 'off-shore' shelters, providing them with havens from national taxation and regulative legislation, and of special 'production zones', 'processing zones', and 'free trade zones' with little or no taxes on value added within national territories where governments assure minimal taxation and regulation. Here, problems arising from 'underregulation' are apparent. A significant part of international business and finance takes place in virtual space: instantaneous electronic transworld communication which escapes all conventional jurisdiction and control mechanisms.

This chapter begins by citing facts and data revealing just how rapid and portentous economic globalisation is. Consequences for economic development policies are then discussed, as well as implications for grass-root innovative and entrepreneurial potential.

Facts and Figures Illustrating the Rapidity of Globalisation

Deep insights into the process, causes and effects of globalisation have been provided, among others but in particular, by S. Sassen, R. Reich, S. Strange, and A. Michalos.[3] Unless otherwise noted, figures cited in the following are from Sassen.[4]

In the 1950s, the major international flow was world trade in raw materials, other primary products, and manufactured goods. Disfunctional international trade and credit practices and policies, which had led to the inability of several developing countries to repay their debts at the end of the 1970s, triggered the accelerated evolution of a new international economic order. In the 1980s, foreign direct investment grew three times faster than export trade and, by the middle of the decade, the main object of foreign direct investment had shifted from raw material extraction and manufacturing to the service sector. Much of the cross-border capital flow involved acquisitions (take-overs) of or mergers with foreign firms, giving birth to transnational corporations. An example of the speed and significance of this phenomenon: by the late 1980s, 80% of U.S. international trade was accounted for by transnational corporations within U.S. borders, one-third of which were actually inter-firm transactions.

American neo-liberal 'financial innovations' (e.g., instruments for securitisation, i.e., transforming financial assets and debts into marketable instruments, for franchising, and for diminishing impacts of changes in interest rates) and development of electronic communication media assured attractiveness and interconnectivity of and instantaneous access to markets for credit, stocks and bonds, facilitating and generating world-wide financial transactions. Their volume exploded: by 1990 stock market capitalisation was equivalent to 64% of Japan's Gross National Product (GNP), to 119% of U.S. GNP, and to 118% of Great Britain's GNP.

Since 1980, the volume of trading in currencies, bonds, and equities has increased five times faster than the Gross Domestic Product (GDP) of all rich industrial countries and the total value of financial assets has increased two and a-half times faster than their aggregate GDP. The foreign currency exchange market is the biggest global market: in 1983, foreign exchange transactions were ten times larger than world trade in commodities. Nine years later, they were sixty times larger! These tendencies are expected to continue and further accelerate. In the age of electronic communication, quoted rates for major currencies can change twenty times per minute. High levels of volatility caused by floating rates have further intensified attractiveness of currency speculation and destabilising capital flows at the cost of production of real goods and services.[5] Due to the degree of concentration in control of capital on the global market, governments and central banks are rapidly losing leverage and their attempts to bail out floundering institutions and to regulate weak or too strong currencies tend to function as a sort of insurance for even more daring 'moral hazard risks' and speculation.[6] Today, there are practically no taxes on this type of international transaction, although a minimal 'Tobin tax', levied comprehensively and uniformly in all countries, would produce enormous revenues for socio-economic welfare and would contribute to a stabilisation of financial markets.[7]

Global financial markets are served not only by banks, frequently transnational corporations themselves provide capital for their own projects and even for those of associates. Capital forming institutions, such as pension funds, insurance companies and specialised investment corporations,[8] are heavily engaged in the global financial market. Not only have global financial markets served transnational corporations and taken over granting credits to developing countries, they have also financed, i.e., 'aided and abetted', huge and irresponsible government deficit spending in the industrialised countries. Between 1945 and 1974, total net public sector debt had dropped to 15% of the GDP in OECD member states. However, since then it has risen to 40%! Thus, global financial markets potentially have a strangle-hold on governments, a potent force in assuring policies for deregulation, low inflation, etc. which are amenable to interests of global capital - without public debates! To cite Sassen:[9]

> Central banks and governments appear now to be increasingly concerned about pleasing the financial markets rather than setting goals for social and economic well-being.

Politicians would, of course, dispute this and claim long-sightedness in their policy-making but long-sightedness implies at least the existence of a long-term concept for taking the initiative in dealing with globalisation instead of simply acquiescing to demands of global capital. It is quite apparent that no major or even secondary political party in any of the major countries has clearly defined a programme for 'taming' globalisation in the interest of social welfare and equity.

The rules of the globalisation game as well as structures and functions of its supranational institutions and participants, be these international agreements (GATT, NAFTA), international agencies (World Trade Organisation, World Bank, International Monetary Fund), or transnational corporations themselves, not only conform to but have been more or less prescribed by Anglo-American neo-liberal economic philosophy.

Consequences for National, Regional, and Rural Economic Policy-Making

Highly integrated but extremely flexible structures of transnational corporations and of financial interest groups operating around the world with multinational staffs and making use of modern electronic communication media as well as of a variety of specialised services have contributed to a rapidly increasing concentration in world-wide control of financial, manufacturing, and commercial markets and, by consequence, in profit appropriation in these sectors. Business decisions on take-overs and mergers have frequently become strategically and economically more important for executives of multinational enterprises than, for instance, decisions on improvements of products and services, on production technology or on sales strategy. In particular with the help of new communication media, global financial interest groups are able to reap enormous profits by free circulation of and speculation in currency, by prompt trading in stocks and bonds, etc. In effect, we are witnessing the rise of a new financial, economic, and political oligarchy, which to an ever larger extent has liberated itself from democratic control mechanisms.

Long before agents of neo-liberalism had access to modern technical means of communication and transportation triggering concentration of economic and financial power, Polanyi[10] stated that uncontrolled market mechanisms would demolish society by destroying or undermining existing institutions of (grassroot) social organisation and, in particular, labour which he regarded as a major constituting element of culture. Ironically, while members of socialist governments around the world were gradually getting hoarse singing the words of 'The International', capital interests began dancing to the tune. But capitalism itself has been transformed in the process. As we have seen in the aforegoing, global capital does not adjust to market mechanisms; it strives to organise the market and to control its mechanisms.

Mainstream neoclassical economic theories and models have almost completely ignored interdependence of both individual and collective economic behaviour with psychological, social-psychological, and socio-political phenomena.[11]

These phenomena are important determinants of organisational processes. Besides its lack of imbrication with other branches of human science, mainstream neoclassical economics has other serious flaws within its own logic with respect to questions relevant to entrepreneurship and innovation. For one thing, as Lazonick[12] points out, mainstream economics has exhibited an obsession with finding *equilibrium solutions* for optimal allocation of market factors (scarce resources), e.g., in response to induced change or in situations where change does not occur. Investment strategy, i.e., postponing consumption in view of higher returns later, does not sufficiently describe processes of resource development (overcoming scarcity by creation of more value/ innovation). Another major deficiency of particular import in the age of globalisation is characterised by what Lazonick has termed the 'myth of the market economy'.[13] Ensuing from changing institutional reality is the fact that 'efficient markets' (or successful business outcomes) are less and less dependent on what Adam

Smith called the 'invisible hand' of supply and demand guiding market forces, but more and more on planned (organisational) coordination of market supply and demand. The 'visible hand'[14] of organisation aims at gaining control over market forces and productive resources. 'Spot markets' - where buyers and sellers independently and competitively pursue self-interests and take action solely on the basis of momentary, objective and impersonal criteria of supply and demand as to quantity, availability, price, quality, need of resources equally accessible to all - have always only existed in the theorist's mind. Even the most simple of contracts involves obligations, constrains market options, and binds resources. This then is what Lazonick[15] describes as a trend from self-co-ordination of the market to *organisational co-ordination* of the market, a trend increasing even more the institutional vectors in economic processes. By gaining privileged access to productive resources organisation can thus supersede 'free' markets. Strategies of organisational market co-ordination determine which activities and socio-economic actors become part of a 'firm' and, on the other hand, which firms amalgamate or join a con-sortium. Coase[16] and especially Williamson[17] attempted to integrate this phenomenon into neoclassical theory with their concept of 'transaction costs' involved with various forms of market transactions. However, even Williamson recognised the existence and strategic importance of 'predatory behaviour' of dominant corporations with the sole objective of bankrupting existing rivals and creating entry barriers to potential competitors.

The following outline of consequences of globalisation on the meso- and micro-economic levels reiterates and expounds upon points raised, in particular by S. Sassen. These points are:
- the loss of state sovereignty,
- disruptive potential of speculation on social and economic development ('casino capitalism'),
- a new valorisation of economic activities entailing new inequalities in income distribution,
- new foundations of citizenship and immigration and
- new economic geography of centrality and siting.

Modern state sovereignty had been based on territorially defined domains of legislative powers and citizenship rights. States were legally empowered to specify explicit civil and social rights and duties of private enterprises and persons as well as to prescribe rules for the circulation of capital, goods, services and, to a certain extent, information. Governments were thus responsible for enacting legislation for social welfare and envi-ronmental protection and, heeding counsel of Keynes and others, made use of a wide spectrum of instruments to regulate the national economy and employment markets, e.g., taxation, public spending, monetary policy, etc.

Faced with economic globalisation, national governments have for obvious reasons, in effect, lost the means to enact and especially to enforce their own laws. In an effort to save places of employment and tax revenues, some state governments have submitted of their own 'free' will to the forces of globalisation and have themselves introduced deregulation measures in favour of free circulation of capital, goods, services, information, and to a limited extent of people (workers) and have provided institutional, material and territorial infrastructure to make their territories more appealing to hypermobile capital and business interests. They have also taken initiatives in instituting supranational legal agencies, agreements and even human rights codes - with varying degrees of effectiveness. Thus, many national governments have willingly transferred certain components of their own authority to 'higher instances'.[18] At the same time, they themselves have undergone transformation in favour of state agencies and policies working for globalisation at the cost of power and prestige of agencies and policies associated with domestic equity and environmental protection.[19] Decline in socio-political participation of the population in economic policy-making is captioned by Sassen's term 'economic citizenship'[20] which now hardly applies to individuals but rather to international firms and global financial interests which can 'vote down' government economic and other policies simply by 'pulling out'.[21]

Casino capitalism

'The global capital market is a mechanism for pricing capital and allocating it to the most profitable opportunity. The search for the most profitable opportunities and the speedup in all trans-actions, including profit taking, potentially contribute to massive distortions in the flow of capital.'[22] These distortions may be even directly counterproductive to economic and social functions of the investment object in question, e.g., to its fulfilment of specific needs for production, services, or employment. This is exemplified by the quote attributed to a leading executive of General Motors: 'We're in the business of making money, not cars'. As a result of distortions on capital markets due to financial speculation and to market organisation (firm merger and acquisition) strategies of investment bankers, SMEs interested in loans for investment in productive capital, in bridging loans or in advance financing of contracts and commissions sense the marked disinterest of banks in giving such loans because their capital would yield much higher returns on global financial markets.

For generations of Marxists, the entrepreneur was regarded as the personification of capital. But it has become all too obvious that capital has become widely dissociated from entrepreneurship. By granting, with-holding, or terminating credit, financial institutions can determine which entrepreneur attains market control. The entrepreneur too feels the strangle-hold of finance. Thus, credit and investment decisions may well be either a matter of pure power politics or of speculation. Even the godfather of state financial stewardship, John Maynard Keynes, recognised that, when finance dominates, 'the development of the country becomes a by-product of the activities of a casino' [23] And it is just the high profit-making capacities of the new growth sectors that makes them a magnet for speculative activity. These processes further the concentration of power in the hands of financial and political oligarchies. They have also led to an upswing in the economies and social welfare of certain underdeveloped countries. On the other hand, Cassen[24] remarks that although one might expect that disintegrating effects of globalisation on highly developed societies, such as a continuous rise in unemployment accompanied by steady declines in real work income and pensions and in the levels of social and health services, would exert pressure on policy makers to reconsider costs and benefits of globalisation, to the contrary, proponents of neo-liberalism proclaim that its principles, i.e., free trade and competition, pri-vatisation, flexibility, priority of 'financial markets' over socio-economic equity, declining role of the state, reduction of state budgets and taxes, cut-backs in social services, fringe benefits of employees, etc., are not being enforced strictly enough.

New valorisation of gainful activity and income inequality

Putting aside for the moment considerations on the huge profit-making potential of financial investment and speculation, it is obvious that multi-national businesses and financial interest groups are in need of highly specialised services and information and have profit-making capacities enabling them to pay exorbitant fees for them. Thus, up- and downstream services to global capital have vastly superior income-generating capabilities as compared to traditional sectors of the economy. This leads to a new valorisation of certain gainful activities with significantly differing impacts on earning capacities within and between the working forces of different sectors. The resulting income redistribution is expected to create new inequalities and socio-economic polarisation.[25]

Precisely, what are the skills most in demand in a globalising economy, i.e., in 'high value' firms specialising in customised services and goods required not only for final consumption but also for intermediate consumption by well-paying transnational corporations? Reich[26] specifies three different but related skills well remunerated by high-value corporations. These are:

- *problem-identifying skills* requiring good knowledge of clients' businesses and needs and enabling the firm to help clients recognise and understand their needs,
- *problem-solving skills* permitting team members to put things together in unique ways and to develop new and individualised applications, combinations, and refinements of services, products and production processes. This is tantamount to innovation.
- *brokering/organising skills* directed at recognising potentials, organising resources, establishing contacts, for instance between problem-identifiers and problem-solvers.

We might note in passing that these are all prerequisites of being an entrepreneur. Reich classifies these skills as 'symbolic-analytic services', the most dynamic and highly paid type of gainful activity. The other two categories in his classification scheme are 'routine production services' consisting of repetitive tasks, and 'in-person services', equally repetitive, but based on person-to-person, hence localised relations.[27] Together, these three categories make up about three-quarters of gainful activities on the American labour market according to Reich's estimate. The rest is composed of those engaged in farming, forestry, and mining or of those working for the government, for government regulated or owned utilities and industries or for state financed programmes.

Management of high-value enterprises tends to contract out or to pay very low salaries for routine and in-person services; at the same time, it squeezes profits for investors. 'Owning' a firm no longer means what it did decades ago. Ever larger shares of corporate budgets go into high fees, salaries, and fringe benefits of the problem- identifiers, solvers and brokers, i.e., to management centres of the corporate network itself. Reich[28] illustrates this by pointing out that, in 1920, 85% of costs in the automobile industry were for paying routine labour and premiums for investors. The remaining 15% went to designers, engineers, stylists, planners, financial specialists, executive officers, lawyers, marketing specialists, etc. By 1990, the allocation to routine labour and to financial capital had sunk to 60%. In modern, 'intelligent' products, for example for semi-conductor chips, costs for routine labour and capital have diminished to 11%; more than 85% of costs being incurred by 'symbolic-analytic' services. It is quite apparent that a view of the working world based on the traditional capitalist vs. labour paradigm is hopelessly antiquated.

Immigration and citizenship

No matter what their individual stance on global circulation of capital, goods, services, and information may be, all nations continue to strive to maintain strict control over the circulation of people by means of immigration laws. Modern transportation infrastructures connecting all corners of the globe provide potentially easy access to wealthy countries for impoverished, underprivileged, or persecuted populations in the 'Third World'. Accords and treaties between nations often make provisions for free circulation of their own nationals within their joint borders of jurisdiction, e.g., the European Union. In addition, nations having acknowledged international codes on human rights are confronted by considerable limitations of their sovereignty, for instance when demands for access to national territory are based on refugee status of applicants, or on petitions for asylum or family reunification. Basically, upon entry to national territories, foreign nationals acquire a wide spectrum of welfare and even civil rights according to prevailing standards in the host country. But even the majority of illegal immigrants do acquire in the long run most, if not all of these rights. However, transnational corporations are also instigators of immigration, in particular of their cosmopolitan staff and highly specialised service personnel, but also of subcontracted or 'guest' labourers. This (at least initially) temporary labour migration is formally accepted, often even promoted by the states involved. The mere existence of large foreign minorities with wide-ranging access to the labour market and to civil and welfare rights certainly does, however, undermine, even devalue advantages of national citizenship.

New economic geography of centrality

Since successful international financial institutions and service industries operate on the basis of quick turnovers in information, they require immediate access to vast networks of personnel, electronic, and physical resources. Messages must flow quickly. Formal scheduled meetings and agendas are secondary; it is the frequent, but informal and spontaneous communications between problem-identifiers, problem-solvers, and strategic brokers which are most productive.[29] By consequence, their headquarters and all 'strategic players' in higher order functions must be located at strategic nodes in these networks. Sassen and others[30] note that, as a rule, firms in more routinised lines of activity with predominant orientation to regional or national markets are free to locate outside of cities. These are most frequently firms engaged in standardised mass production of tradable goods and lower-end services employing older technologies and lower skill labour. In such cases, advances in telecommunications may indeed contribute to a decentralisation of economic activity in space.[31] On the other hand, multinational enterprises engaged in highly competitive and innovative activity with an orientation to the world market locate themselves, by necessity, in cities providing them with the resources they need while not binding them to any specific location, be it the city itself or the region. Flexible, mobile, high-value enterprises do not want to be weighed down with 'overheads', such as office buildings, plants, equipment, and payroll staff.[32] Speed and agility in 'wheeling and dealing' are important in a business sector in which capability of cleaving linkages quickly may be just as essential as swift initiation of new ones. Hence, in high-value enterprises there are few people who have 'steady jobs' with fixed salaries. Instead, collaborators tend to have temporary or project-bound contracts granting high levels of income and a share in risks and returns.[33]

High incomes of key corporate personnel and the resulting rises in costs of rent and services tend to push less dynamic or financially potent enterprises and segments of the population out of central locations. Sassen[34] points to the formation of a new transnational urban system. This network of 'global cities' is not only of logistical importance as a system of financial pivotal points and firm headquarters with their respective service agencies but is also a 'freeway' for the trans-nationalisation of labour and the formation of cosmopolitan identities.

Can socio-economic impacts of globalisation be spatially differentiated? In particular, what will happen in rural areas? Aside from the descriptive analysis by Sassen and others on global cities, their urbanised surroundings, and 'spatial-spillovers',[35] few results are available for regionally differentiated analysis of effects of globalisation that are not restricted to case studies. Regional economic forecasts have much in common with meteorological forecasts: The main component forces are known but their specific interaction within highly varied regional and rural contexts is rather difficult to predict. The general validity of findings for their application to very heterogeneous circumstances found in individual rural areas will always be questionable. Nevertheless, general consensus on probable socio-economic effects of globalisation on rural areas does exist, even if some data are inconclusive and expectations have not always been confirmed to date.[36]

Observations indicate that, in general, non-major cities and vast rural areas outside of global agglomerations are becoming increasingly peripheral. Labour markets are hardly influenced any more by agricultural policies and even up- and downstream enterprises serving the agricultural sector contribute little to employment in most regions and are themselves also subject to mergers or acquisitions.[37] These trends engender new spatial patterns of economic dynamics, income distribution and purchasing power of the population cutting across rich as well as poor and developing countries alike. OECD data on effects of globalisation on rural areas are ambiguous.[38] The fact that rural areas in the highly industrialised OECD countries produce a disproportionately high share of tradable, mass produced goods necessitating vulgarised and low skill technologies makes them vulnerable to relocation of production to countries with cheap labour. Likewise, the service sector has grown in rural areas, but it concentrates on lower-end, consumptive services. Due to low densities of the population and enterprises, rural areas can hardly offer the wide range or critical mass of highly specialised services necessitated by modern lean management of transnational corporations. Positive agglomeration effects on exchange and transfer of specialised know-how between enterprises and individuals are lacking in rural areas, making considerable efforts necessary to take advantage of recent developments in technology. Rural areas have, with a few notable exceptions like Silicon Valley, been the end station of the technology diffusion process. In today's global capital market, the pool of regionally accumulated capital in rural areas, which otherwise might have remained for investment in the region, may depart for who-knows-where. Certainly most investors in certificates of deposit and investment funds have no idea exactly where there money is being used. The tendency is evident: Capital will tend to flow from lagging areas to wealthy, dynamic areas, like global cities, where higher rates of return can be expected.[39] Rates of return on regional capital markets formerly considered to be quite good cannot compete with very high rates coupled with comprehensive services offered by agents on the global capital market.

In any event, Freshwater[40] finds that in a globalising economic framework a shift in importance for siting is taking place from 'comparative advantages' of locations, e.g., relative advantage with respect for endowments with natural resources (traditionally an asset of rural areas) to 'competitive advantages' in social and political arenas. There is considerable evidence that small urban centres in rural areas of certain highly developed countries like Germany and the Netherlands have been well equipped with both advantages because they have been able to offer sufficient agglomeration and synergy effects to maintain or even improve their relative status within their national economies and are viewed by some to be the pillar on which rural development must rely.[41] Development policies should thus aim at reinforcing remaining competitive advantages of rural centres and at improving infrastructure and technological skills of the population. It follows that cultivating innovative and enterprising predispositions of the population as well as establishing amenable socio-cultural environments for innovation and enterprise must be accorded high priority on the policy agenda.

Although Sassen has, *pro forma*, acknowledged some major drawbacks of globalisation, in our estimation, she has neglected 'the human side of the story' in concentrating on formal analysis of globalisation while ignoring the fact that behind formal structures and relationships there are social contexts[42] and, above all, people: People who are amassing power, building networks, and lobbying. Globalisation takes place in a carefully spun net of interpersonal relationships which tie officers of financial consortia, politicians, state administrators, managing executives, brokers, consultants, and other highly qualified service personnel together. Their decisions often transcend legal and often economic frameworks (which can be later rectified by legal consultants, etc.) and are frequently based primarily on mutual trust between protagonists. These relationships require more than 'virtual' proximity (i.e. contact via electronic media); face-to-face relations have not lost significance. Closely tied into these networks are great numbers of politicians, many of them formally as either incumbents of seats on boards of directors/trustees or as consultants, as well as countless state administrators who devise and implement deregulation measures and regional development policies involving subsidisation or other policies favouring corporate interests. Finally, we have those who officially monitor and evaluate these measures. As a rule, these 'neutral arbitrators' are themselves part of the 'web' or are at least dependent on the 'web'.

Facts presented above make perfectly clear: If macro-/national economies are to provide material subsistence and productive activity to the population as a whole (as the German term *Volkswirtschaft* would indicate) globalisation will have to be tamed. *Sustainability of regional development is certainly not on the global agenda.*

Differentiating Target Groups as Agents of Regional Development

Promoting self-employment in general can alleviate unemployment only to a small extent. Regional development policy should thus *differentiate more clearly between various other forms of self-employment,* on one hand, *and owning and managing a business enterprise* on the other. As a rule, the self-employed person in a classical liberal profession (e.g., medical practitioner, lawyer, notary public, psychologist) is him- or herself the person occupying the major and often sole job, all other activities being auxiliary. The commercial sector (e.g., retailing) has proven to be an important training ground for new entrepreneurs[43] and is relatively easily accessible to newcomers. However, the 'turn-over' (establishment and dissolution) of commercial enterprises is, in the rhythm of economic cycles, relatively constant and does not have a great job-creating potential. Classical trades and handicrafts (e.g., plumber, carpenter, mechanic, electrician/electronic engineer) are indeed excellent 'incubators' for new businesses and do have a small job-creating potential. But, according to the analyses cited above by Sassen and Reich, trades and handicrafts, wholesale and retailing, and classical liberal professions are mostly routine/repetitive, in-person i.e. consumer-oriented gainful activities with relatively low innovative potential and, for the most part, destined to decrease in their relative earning potentials. They also have a very limited action radius. Furthermore, self-employment in a liberal profession, in the commercial sector, or in a handicraft requires some but not all skills and functions involved in entrepreneurship in the sense of owning and operating a business and have a somewhat different motivational basis of action. All entrepreneurs are self-employed but not all self-employed persons are really entrepreneurs. And many people responsible for business operations are neither self-employed nor owners.

It was only in the late 19th and early 20th Centuries that 'managerial capitalism', as opposed to the traditional personal union of owner and manager in 'personal or proprietary capitalism', emerged as a wide-spread form of directorship.[44] It is clear from preceding sections that today's globalising economy represents the heyday of managing executives of financial consortia and of transnational corporations who have managed to shake off government control and democratic censure. In fact, they are even in a position to strongly manipulate most stockholders and institutional investors and thus to limit the influence and demands of capital interests. Again: company 'ownership' no longer means what it once did. Neither does company leadership. In the webs of transnational corporations patterns of leadership are constantly being reconfigured and new leaders emerge due to their reputations and informal power coalitions. Ownership and 'control' of high-value enterprises is usually diffuse and not, for a considerable duration, in the hand of a single individual or group.[45] The key asset of high-value enterprises is neither real estate nor material assets but rather human capital which - we repeat - through accumulation of experience is the only kind of capital capable of rendering increasing, not diminishing returns. Corporate management aiming at providing customised goods and services to clients in rapidly changing markets realises that the profile of skills and expertise it needs changes from day to day. This means that flexible staff configuration must be possible. Hence, due to high profitability of temporally limited use of appropriate human capital and due to externalisation of responsibility for social welfare and further professional training of its staff, corporate management is willing to pay exorbitant salaries to and to share risks and returns with those collaborators for specified periods of time or for specified projects in strategic-analytic tasks. Routine work is destined to be contracted out.

Our close attention is merited by management personnel of such transnational, 'high-value' corporations since their activity profile and the innovative and linkage skills required by their functions resemble in many respects those of the 'classical entrepreneur'. It is the author's contention, however, that personality traits, personal goals, and certain key functions of members of these corporate management networks differ in important ways from those of entrepreneurs who we consider to be grass-roots agents of innovation and enterprise, namely owner-managers of small or medium-sized enterprises (henceforth called SME-entrepreneurs). In our estimation, policies for regional development targeting corporate management would, by consequence, seem to be mainly geared to augmenting personal roles, goal attainment, and power of members of corporate networks rather than to cultivating sustainable local and regional human potential for innovation and entrepreneurship. The following schematic comparison elucidates some essential differences between these two groups, corporate managers and SME-entrepreneurs, otherwise sharing many similar traits.

Table 2.1 Basic differences between corporate executives and SME entrepreneurs

CORPORATE EXECUTIVES	SME - ENTREPRENEURS
• (incumbent leaders/ 'managerial capitalists'	• irreplaceable *leaders/* 'proprietary capitalists'
• *dependently employed* subject to critical evaluation	• *self-employed* self-*responsible*
• contractually *fixed salary* possibly with profit sharing, stock options	• *direct profit appropriation*
• involved in *cosmopolitan information networks*	• involved in *cosmopolitan information networks*
and is a	*but* is a
• *global player* with a high degree of *personal mobility*	• *non-global player* with *durable ties to the region* incl. *business* overheads
• attitudes influenced by self-identity as member of a *global caste*	• attitudes influenced by a *strong local and regional identity*
• non-membership of local/ regional communities as a defining trait	• prestigious and personally engaged *member* of local community
• actions characterised by *weak identification with product / service* and by	• actions governed by *strong identification with innovative product/service* and by
• *strong interest in commercial and financial spot markets*	• *strong interest in reputation and durable working relationships*
• *no or strongly mitigated risks* to own private assets in case of business failure but generous provisions for severance	• *high degree of risk* to own private assets in case of business failure

Obviously the SME-entrepreneur is irreplaceable as owner and ultimate authority in his or her firm, whereas the corporate executive is merely the incumbent of a position in the firm. As opposed to our classical entrepreneur, the corporate officer's position depends on continued support of others. The corporate officer receives a steady salary, sometimes with fringe benefits coupled to the firm's success. The firm's profits are at the full disposition of the entrepreneur, however.

Within a global framework, potentially successful SME-entrepreneurs must be just as cosmopolitan in their scope of perception and in the reach of their socio-professional networks as managers of transnational corporations. This is imperative not only for achieving and maintaining current levels of technological know-how but also for exceeding them (innovating). It is also a prerequisite for recognising and winning new markets on the national and international scale. However, in contrast to the corporate manager, the *SME-entrepreneur remains essentially a non-global player,* which is exemplified by the fact that the firm's *production base tends to remain fixed in the region* - frequently the region of his or her origin. These ties are not only manifested in emotional affiliation or in intimate knowledge of specific regional potentials and hazards but in the willingness to take on and deal with 'overheads'. Whereas 'flexible' management of dynamic transnational corporations desires to remain 'weightless' and mobile and thus does not want to have its capital assets tied down by ownership of real estate and machines or long-term work contracts, the SME-entrepreneur continually establishes or renews the firm's ties to the region, the population, and its political and economic institutions by assuming 'overheads' in the form of investment in location-bound production factors, such as real estate, equipment, permanent or long-term contracts with core personnel and by cultivating long-lasting relationships of trust and collaboration with local customers, business partners, and representatives of local business and public institutions. As opposed to corporate headquarters, the SME-entrepreneur is inclined to provide steady jobs with fixed salaries to a certain number of trusted and reliable collaborators in traditional, often preferably rural locations.[46] For corporate managers the ability to abruptly sever relationships and to terminate contracts may be more vital than establishing new ones. This also applies to their own work contracts; corporate officers thus must retain a high capacity for personal mobility and detachment from the local community. The economic behaviour of SME-entrepreneurs is much less oriented to so-called 'spot markets' where transactions are determined by a number of 'objective' market parameters like price, quality, availability of services and goods. It is much more strongly conditioned by other unobservable and immeasurable factors, such as implication of one transaction for a series of future transactions or conclusions from past experience relative to reliability, honesty, solvency of business partners and customers.

The above-quoted statement of the General Motors' executive about being in the business of making money and not cars is revealing about corporate management's degree of identification with the product or service in question. Marketing 'no-name' products and liberal franchising of known trade-marks (selling brand names and reputations) are other unmistakable indications of corporate management's detached relationship to the corporation's output. 'Flexible' corporate managers are constantly on the look-out for the most lucrative capital spot market, regardless of the momentary area of production or service in which they are engaged. By contrast, successful *SME-entrepreneurs* ('hidden champions') have been shown to *specialise and to highly identify themselves and the firm with the quality and reputation of their special product, technology, or service.* [47] Attention and resource allocation are concentrated on domination of related segments of regional, national, and global markets. A related differentiating factor between SME-entrepreneurs and corporate managers is the *degree of coincidence (unity) of personal and firm identity.* As we have just mentioned in the aforegoing, corporate managers are by their institutional status merely incumbents of their positions and are thus subject to replacement from someone from within or without the firm, not necessarily having further-reaching repercussions for the firm's operations. Consequently, 'flexible' corporate managers would generally be willing to accept a more lucrative opportunity, even from the firm's 'worst' competitor. As a result, the degree of coincidence (unity) of personal and firm identities is minimal. The *SME-entrepreneur is the firm and has a vital interest in its success and continuity.* By his or her very role and self-identity he or she is indispensable and the ultimate instance in all decision-making. The firm's existence and destiny are strongly linked to the life-span of the SME-entrepreneur and to provisions made for intergenerational transfer, take-over, amalgamation, joint ventures. Hence, this does not preclude the entrepreneur's strategic involvement or even a long-term inclusion of his or her enterprise in a corporate 'quasi-firm', in fact, this type of collaboration may prove to be a welcome 'division of labour' between contracting parties, providing corporate management with sub--contractual access to material and manpower on one hand and opening new markets to the entrepreneur on the other. Nevertheless, the entrepreneur will not accede to a take-over and will retain in the long run overall personal autonomy of his 'module' in the 'quasi-firm'. [48]

SME-entrepreneurs and corporate managers also differ in the kind and amount of personal risk involved in the business. Although success of collaborators may be rewarded by supplements to salaries or even by some degree of profit-sharing, in general, firm ownership still means to the SME-entrepreneur that, having assumed the risks, he or she is entitled to most profits. The *SME-entrepreneur stands to lose all business assets which are ultimately (the bulk of) the family's own private assets.* The risk of corporate managers is, at best, strongly mitigated, even if they have stock options or have invested some of their own personal assets in the firm. Their main security (capital) is the reputation and professional net-works they have built up in the course of their management which they can take with them, if need be. Fringe benefits provide security (e.g., pension and health insurance) which, under certain circumstances may have greater longevity than the firm itself. In addition, corporate managers are adept at making contractual provisions for severance from the corporation in the form of 'golden parachutes' consisting of lucrative settlements for severance allowances, annuities, pensions, etc. which legally supersede property rights of owners or stockholders. Managers, particularly in large corporations with numerous stockholders, have wide leverage and are capable of pursuing their own interests at the cost of stockholders. This phenomenon can be observed in both Western and formerly socialist economies.[49]

It is apparent that the *SME-entrepreneur, through longer-term relation-ships to localised assets including human capital, is a verified member of local and regional communities.* This is not the case of corporate managers in transnational, high-value enterprises, in fact, non-membership would more likely be a defining characteristic of this group. Lee[50] comments that globalisation tends to diminish social solidarity by reducing civic engagement of such internationally mobile groups. Although at the centre of increasing concentration of economic power, personal leadership in transnational corporations is diffuse and revocable, while the (SME-) entrepreneur in our sense leaves no doubt who is in command and that the firm will persist for many years, if not for generations. The *SME-entrepreneur stands out as the dominant individual in a hierarchical structure* whereas the corporate leader is simply the current incumbent in an ephemeral organisational configuration and the incarnation of interplay within personal and organisational networks. *The SME-entrepreneur takes full risks and wants full returns. It is more likely that profits he succeeds in making will remain in the region.* For these reasons we see the traditional entrepreneur (owner and director of a SME) as being the primary target group for regional development policy. In our view, it is this type of individual whose interaction with the socio-economic environment is most promising for initiating and sustaining innovative and entrepreneurial activity in a given region. These are the individuals to be targeted in measures to strengthen regional development potentials.

Individuals do not, in our estimation, drift coincidentally into one or the other of these socio-economic roles but have important personal predispositions. The role of a SME-entrepreneur certainly requires an enterprising, innovative, and independent personality while the corporate officer exhibits many bureaucratic characteristics. We shall illustrate our point by following up the above comparison of SME-entrepreneurs with corporate executives with a scheme in Table 2.2 juxtaposing two 'phenotypes':

- 'bureaucratic, organisational man' [51] and
- 'enterprising, innovative, and independent man'.

In outlining significant differences between these types in the following scheme, reference will be made in part to a differentiation by Bierstedt[52] between characteristics of 'formally' and 'informally' organised groups. We have largely expanded upon this typology, however. Some of the characteristics applied by Bierstedt to 'formal organisations' can, in modified form, be attributed to 'organisational man'. In spite of many traits and functions shared with SME-entrepreneurs ('enterprising, innovative, and independent man') corporate management personnel basically correspond to 'bureaucratic, organisational man'. The same can be said about the association of 'informal organisations' with 'enterprising, innovative, and independent man' and formal institutions with 'bureaucratic, organisational man'. In our view, *it is the SME-entrepreneur who personifies the 'entrepreneur' described in the following chapters* and whose interaction with the socio-economic environment is most promising for sustaining innovative and entrepreneurial potentials in a given region. *This is the target group for measures to strengthen regional development potentials.*

By constructing a 'soft' typology of 'bureaucratic, organisational man' as opposed to 'enterprising, innovative, independent man' and by stressing potential conflicts of disposition and interests between these differing personality types, we have endeavoured to accentuate the significance of better formulation of regional development policies with long-term objectives according to needs of, in our estimation, the most promising target groups at the grass-root level: SME-entrepreneurs and potential initiators of new ventures. An unwanted, but almost inevitable side-effect of this approach would be to leave the impression that these two groups be natural opponents to one another. Nothing could be further from the truth. In fact, these groups depend on one another; each has its role to play in regional development. SME-entrepreneurs must be able to rely on good working relationships with both global corporations and capital interests, on one side, and with local administrators and politicians, on the other. Notwithstanding, it would appear regional development policies presently perpetuate an advantage for organisational types.

Table 2.2   Typology juxtaposing two personality types

BUREAUCRATIC, ORGANISATIONAL MAN	ENTERPRISING, INNOVATIVE, INDEPENDENT MAN
• *orientation to*: highly formalised norms	• *orientation to*: individualised cultural norms
• *behavioural determinant*: status	• *behavioural determinant*: role
• *reward:* social prestige	• *reward*: general esteem, firm profits
• *basis*: conferred authority	• *basis*: personal leadership, ownership
• *bilateral rapport*: super- or subordination, temporary relationships	• *bilateral rapport*: dominance or submission, durable relationships
• *evaluation of persons*: extrinsic (trust in authority or position)	• *evaluation of persons*: intrinsic (trust in individuals, personalities)
• *type of interaction*: status relations, cooperation	• *type of interaction*: personal relations, competition, sometimes cooperation
• *personal inclination*: personal security, following path signs	• *personal inclination*: risk-taking, making one's own path
• *priority values*: conformity, obedience, adherence to customs, organisational affiliation, personal success and reputation in role	• *priority values*: individuality, personal freedom, deviation from customs, independence, success on market, good reputation of firm and output
• *type of activity*: regulation/control	• *type of activity*: competition
• *economic function*: management administration, (re-)distribution	• *economic function*: productivity/ innovation
• *personal goal*: self- and group representation	• *personal goal*: selling ideas, products, services
• *primacy of* means, procedures, peer approval	• *primacy of* goals, success, being first/ best,

It has been argued (by state-employed 'farm scientists' in Germany) that the presented typologies of 'enterprising, innovative, independent man' and 'SME-entrepreneurs' not be applicable to 'European man' but rather to 'American man'. The author's contention is precisely that shackles laid on entrepreneurial personalities by dependency culture and a strong bureaucratic, oligarchical environment in Europe really do strongly inhibit metamorphosis of a large part of basic human potential into dynamic and innovative and enterprising behaviour. Indeed, increasing 'organisation' of markets is also putting pressure on the innovative entrepreneur in North America as well. Nevertheless, basic human potentials do remain, even if somewhat dormant, and can be activated if political and socio-cultural circumstances permit. This then is the *Leitidee* in what follows.

Notes

[1] Strange 1986. p. 1f.

[2] Cf. Rodwin and Sazanami 1991, p. 30f.

[3] Cf. Sassen 1996a, 1996b, and 1994; Reich 1991; Strange 1986; and Michalos 1997. Cf. also Lazonick 1991; Kreps 1990; and Chandler 1984.

[4] Sassen 1996a and 1994.

[5] Cf. Michalos 1997, p. 12 and 32.

[6] Cf. Michalos 1997, p. 27ff.

[7] An excellent presentation and evaluation of arguments for and against a 'Tobin tax' can be found in Michalos 1997.

[8] Sassen notes that 'institutional investors' manage about two-fifths of private household assets in the USA, up from one-fifth in 1980.

[9] Sassen 1996a, p. 50.

[10] Polanyi, 1944, p. 72f; see also Glasman in Bryant and Mokrzyeki 1994.

[11] An excellent, recent review of the relationship of economics to other human sciences can be found in: Lewin 1996. Lazonick commented that champions of mainstream economics make a practice out of evaluating reality using neoclassical theory instead of vice-versa, cf. Lazonick 1991, p. 6.

[12] Lazonick 1991, pp. 10 and 64ff.

[13] Lazonick 1991, e.g. p. 16.

[14] Chandler 1977 and Lazonick 1991, pp. 7, 11, 23ff, and 59ff.

[15] Lazonick 1991, pp. 8, and 191.

[16] Coase 1937.

[17] Williamson 1985; See also Lazonick 1991, pp. 128 and 373ff.

[18] Cf. also Chesnais 1993. Irmen und Blach (1996, p. 718) see evidence of remaining leverage of national policies for economic governance exemplified in positive effects of German policies for rural development and of liberal Dutch policies for the employment market. This is, in our estimation, an overly optimistic view of the government's leverage. Certainly, dense and modern transportation and communication infrastructure in these countries have made possible certain spill-over effects from global cities.

[19] Sassen 1996b, p. 214.

[20]   Sassen 1996a, p. 38ff. The intellectual astuteness which Sassen demonstrates in her dissection of the globalisation phenomenon stands in sharp contrast to her rather naive allusions to new bases of democratic participation in global economic pro-cesses, e.g. stockholders and financial investors 'voting with their feet' (as if even most members of these social minorities would be able to closely and continuously monitor use of their capital by business and financial management) and to the significance of supranational 'human rights codes' in transactions of international business and finance.

[21]   Or by declaring sudden scarcity of commodities, raw materials or energy (oil), as former US-President Carter could testify.

[22]   Sassen 1996a, p. 51. Also see, in particular: Strange 1986.

[23]   Keynes cited by Sassen 1996a, p. 42.

[24]   Cassen, B.: Dienerin statt Herrin. Die Zeit (38) 12. 9. 1997, p. 39.

[25]   For instance, Radermacher (1997, p. 9) assesses a 20%/80% ratio of winners to losers in this process. See also Michalos (1997, p. 31) who notes that more than half of US stocks are held by the upper 1% of the population according to wealth distribution in the USA.

[26]   Reich 1991, p. 84ff.

[27]   Cf. Reich 1991, p. 174ff.

[28]   Reich 1991, p. 104f.

[29]   Cf. Reich 1991, p. 88.

[30]   Sassen 1996a, p. 11. See also Reich 1991; Irmen und Blach 1996, p.714; and Freshwater 1996, p.779f.

[31]   A study by Black, Bryden and Sproull among enterprises in the Scottish Highlands and Islands showed that the overwhelming majority of grass-root respondents found that the introduction of telematics into their area
        - improved speed of communications:
        - helped sink administrative, travel and marketing costs
        - widened their firm's selection of products, services, and activities
        - enhanced their skills
        - improved access to data and information
        - saved time
        -increased opportunities for 'distance working'.
        Cf. Black, Bryden, and Sproull 1996.

[32]   Cf. Reich 1991, p. 89ff.

[33]   Cf. Reich 1991, p. 90f.

[34]   Sassen 1996b, p. 212ff; see also Rodwin and Sazanami 1991, p. 23.

[35]   Cf. Quah 1996.

[36]   For instance, expected negative impacts on (in international comparison densely populated, highly interconnected) 'rural areas' in Germany are not yet recognisable in various indicators. German rural areas are not yet to be considered 'losers' of global-isation. Cf. Irmen und Blach 1996, p. 714 and 718, and von Meyer, 1996, p. 738ff.

[37]   Cf. von Meyer 1996, p. 740.

[38]   Cf. Freshwater 1996, p. 779ff and von Meyer 1996, p. 733ff.

[39]   Freshwater (1996, p. 778) cites Gunnar Myrdal's concept of two counteracting forces, i.e. on one hand, of an equalising spread of resources in market economies contribut-

ing to a convergence of different locations with respect to assets in labour, capital, and products (entropy) and, on the other, 'backwash' of these resources to wealthier areas in quest of higher rates of return. Cf. also Higgins and Savoie 1995.

[40] Freshwater 1996, p. 777.

[41] Cf. Irmen und Blach 1996, p. 716; Simon 1996, p.10; and 719 and von Meyer 1996, p. 734.

[42] Cf. e.g. Mingione 1991, p. 51ff.

[43] This was found to be particularly the case in economies in transition, cf. Roberts, Adibekian et al. 1997, p. 11.

[44] Cf. Chandler 1984, p. 473. See also Lazonick 1991, p. 23ff

[45] Cf. Reich 1991, p. 98f.

[46] Simon 1996, p. 10.

[47] Cf. Simon 1996.

[48] See Simon 1996, in particular Tab. 4, p. 11.

[49] Feldmann 1997, p. 301ff. Often, premiums or stock options are offered to managers as an incentive to pursue mamagement policies leading to rising stock values. Rising stock values are not necessarily an indicator of firm efficiency or innovativeness; they may reflect, for instance, an increased market segment or control.

[50] Lee 1997, p. 182.

[51] Simon speaks of 'administrative man' as a brother (or counterpart) of 'economic man' but this type would encompass only part of what we shall term 'bureaucratic, organisational man'. Cf. Simon 1957, p. 241.

[52] Bierstedt 1957, p. 315ff.

3 Behavioural Dynamics at the Grass-Root Level:

Innovation and Entrepreneurship as Functions of Personality, Social Structures and Cultural Configurations

A social entity composed, e.g., of bees or ants would under controlled circumstances always react in a foreseeable manner. However, human societies are unique insofar as they are not composed of simply reproduceable and interchangeable members but are rather a collectivity of individual personalities. Hagen[1] states that change in society as a whole will not occur without change in personalities.

In analysis of dynamics in economic and social behaviour at the grassroot level, some (non-mainstream) economists[2] and sociologists[3] have long realised that concepts of 'atomistic action' by firms or households are completely unrealistic and that particularities of individual actors and of their embeddedness in socio-economic structures inside and outside these collective entities are important determinants of firm and household 'behaviour' (i.e., actions). Lerner[4] saw economic growth as a function of the significant presence of 'mobile (flexible) personalities' whose values and cognitive processes are strongly based on rationality and openness to novel ideas. McClelland[5] emphasised the significance of the 'entrepreneurial spirit' in personalities for the growth of the economy, and even of civilisation itself. He considered it to be much more important than external resources or actions by planners and politicians. We shall later deal with the indicator he chose for entrepreneurial spirit - the personality trait: 'need for achievement'. It is at this point particularly noteworthy, however, that his historical investigation (e.g. using literature analysis) was able to identify periods of intensified general interest in 'achievement' which were always followed by dynamic economic growth.

In this chapter we shall point to *synergies between human capital and non-economic factors stimulating regional incubation of innovation and entrepreneurship.* In effect, this chapter provides the argumentational foundation for two subsequent chapters dealing in depth with the significance of individuality of social actors (3.1, 3.2 and 3.3 as heralds of Chapter Four) and with the significance of socio-cultural contexts in which the actor is embedded (3.4 and 3.5 as heralds of Chapter Five).

48

Genesis of Enterprise and Innovation

What do we know about new firm births and about the pool of potential entrepreneurs in the general population? Most empirical insights have been provided by research in the U.S.A.

The 'population ecology model' of organisational development focuses on significance of social and cultural environments for the formation and transformation of organisations, specifically with respect to environmental capacity for making available or for withholding resources.[6] While large corporations rarely, if ever, go out of business but are usually taken over by other firms or merge with them, Aldrich[7] points out that about half of all small businesses fail within two years of their creation and probably less than 20% of them ever achieve anything resembling 'success'. Some types of firms have relatively constant birth rates and turnover, other types of business foundations are seldom or sporadic. In addition, organisations sometimes radically change their structure and processes in the course of time to become something completely unforeseen at their time of creation. Aldrich[8] enumerates a number of environmental factors increasing emergence or transformation of organisations:

- variability of the population depending on exposure to ideas from other societies or regions, on attitudes based on secularism and belief in science, and on quality of communication and transportation networks at its disposal,
- shifts in environmental selection criteria due to changes in the distribution of resources and in terms of trade in their exchange, opening new niches,
- the breakdown of retention mechanisms protecting old organisations from competition.

These situational variables in organisational population ecology models are largely manifestations of regional behavioural dynamics.

Reynolds[9] quoted two studies indicating that two-thirds of those starting new firms were doing so for the first time; of the remaining third, half were doing so for the second time and the other half for the third or more time. These studies involved respondents who had actually already taken some action to initiate a new venture. As for simple expressions of intent, Reynolds cites a telephone interview of a 'representative' sample of ca. 1200 adult individuals: 34% reported that they had already been involved in setting up a firm sometime in their lives; 13% stated that they were thinking seriously about starting a new business; of those seriously considering firm set-ups 24% had actually taken some steps nine months later. This amounted to 3% of the initial sample - more than would have been anticipated on the basis of past analysis. There was a consistent disparity between men and women in both past experience in entre-preneurship and in the proportion of the respective subsamples thinking seriously about firm set-ups. Findings commonly also indicate that firm initiators usually do so in familiar locations and in the same economic sector where they have gained relevant knowledge and contacts through professional experience. Thus, these findings indicate that regional pools of entrepreneurship potential are larger than generally assumed and that there is a preponderance of males. They also illustrate the significance of incubator organisations for conferring basic know-how and experience and of the local or regional context for business set-ups.

Rapid development at the regional level depends on the swift flow of information, fast diffusion of innovations and quick acceptance of new ideas.[10] *Innovation* is the product of mental activity combining two or more elements that had previously not been joined in this fashion; innovations are thus configurations of pre-existing components.[11] Stöhr[12] defines 'innovation' as the first commercial utilisation of new scientific-technical knowledge within one enterprise, which he further differentiates as either 'product innovation' or 'process innovation'. High technology enterprises are pragmatically defined by Stöhr as firms with an above average ratio of Research and Development expenditures to net sales, an above average proportion of employees engaged in engineering, scientific, professional, and technical work, and rapid growth in terms of employment and output. For Hagen[13],

> innovation, understood to be both creative mental activity and conversion of novel concepts into new actions or material forms, is Step 1 in entrepreneurship. Step 2 is personnel, financial, and organisational management and opening up of markets.

Rogers and Shoemaker's[14] analysis of receptors in innovation dif-fusion processes differentiates the following adopter categories according to the point in time of adoption: *innovators*, *early adopters*, *early majority*, *late majority*, and *laggards*. Research on 'adopter' categories has looked for and found several relationships between the time of adoption (relative to the general environment) and various social, economic, demographic and locational characteristics.[15] Analysis of the temporal dimension of the cumulative proportion of adoption in a given society reveals an S-curve function depicting an initial slow, but gradually increasing number of adopters, followed by a phase of rapidly increasing numbers of adopters, gradually tapering off as the few remaining pockets of resistance to the innovation disappear.[16] Arthur, Katz and Shapiro, and Dalle et Foray[17] all point to the fact that for many types of innovations benefits accrue the most at later rather than at initial stages of the innovation process, e.g., having a fax machine is most advantageous when their use is widespread and there are many potential senders and receivers.

According to Camagni[18] there is a fairly general consensus about *characteristics of firms* which are most likely to rapidly adopt innovations. These features pertain to the firm's age, sector of activity, management style (e.g., weak hierarchy, flexibility in delegation of responsibility, receptivity for information, minimisation of routine activities), and its specific mix of research and marketing activity. In the age of high technology, importance of 'technological apprenticeship' for formation of technologically versatile and innovative human capital is quite apparent.[19]

Cochran[20] bemoans a lack of primarily social models of economic change, specifically of models including the entrepreneur's and innovator's role. Unfortunately however, there is no universally accepted definition of the 'entrepreneur'.[21] Gartner offers a comprehensive synopsis of the numerous definitions of the entrepreneur in scientific literature.[22] Low and MacMillan[23] also cite a number of definitions and typologies of various kinds of entrepreneur. It was Cantillon who first coined the term '*entrepreneur*'.[24] Casson[25] and a number of other authors claim that John Stuart Mill introduced the term in England. Classical economical theorists such as Adam Smith applied it to individuals assuming certain economic functions, namely perceiving market opportunities; subsequently obtaining, combining, and using resources to produce goods and services satisfying market demands, and finally marketing the products. Say[26] defined the entrepreneur's role as combining resources for production of goods and services. These were basically definitions of functions (of entrepreneurship), not of types or traits of individuals (of entrepreneur types). However, there have long been considerable efforts, particularly in the field of psychology and social psychology, to personalise the entrepreneur by identifying defining traits. Many have gone so far as to claim: 'it's in the genes'.[27]

Defining 'Entrepreneurs' (and 'Innovators') by Traits

As far back as 1803, the above cited J.B. Say had claimed that entrepreneurs must possess a scarce combination of (moral) qualities such as intelligence, prudence, integrity and reliability. But it was Sir Francis Galton (1822-1911) who pioneered empirical trait research in individuals in general. Mischel[28] specifies that 'traits' have widely been regarded as a construct accounting for behavioural consistencies and differences, being incorporated in persons and determinative of predisposition to certain types of behaviour. He later proceeds, however, to demonstrate that traits have a poor record in behaviour prediction since the individual receives a multitude of behavioural stimuli from various, equally influential situations and environments. In spite of these reservations he notes that some traits may be transsituationally more persistent than others, e.g., intelligence shows some stability and generality in its effect on achievement-related behaviour.

> Nevertheless, even indices of intellectual performance and cognitive abilities and skills may be more situation specific than trait theorists initially assumed.[29]

Vesper[30] identified 12 'barriers' to entrepreneurship, including personal attributes such as a lack of market knowledge, inability to delegate responsibility, and a lack of technical skills. The question remains to be pursued in subsequent chapters how the appearance of, possibly dormant, more or less favourable traits is influenced by situational contexts.

Defining 'Entrepreneurs' (and 'Innovators') by Functions

In the definition and study of entrepreneurship, a dispute between those highlighting 'traits' of entrepreneurs, those stressing their 'functions', and those emphasising the importance of situational opportunities has persisted over decades.[31] For instance, Gartner[32] and Jenks[33] emphasised that studies should focus on what the entrepreneur does, not on who he or she is. There are a number of economic roles/functions which are largely but not totally synonymous with entrepreneurship, such as self-employment and management, and it is not always easy to differentiate them at all times and under all circumstances. Brockhaus[34] cites the current definition of the 'entrepreneur' as one who initiates a venture, organises, owns, manages, and assumes risks of the business. 'Self-employed individuals' are commonly defined as earning no wage or salary but deriving income from a profession or business on their own account and risk. Entrepreneurs are identified as a subset of this group, sometimes with the additional criterium of being the sole owner of the enterprise.[35] As we have seen in the above, 'corporate managers' assume a variety of functions similar to the entrepreneur in the acquisition and allocation of resources but did not initiate the venture, do not own the business, and do not assume, to a comparable degree, risks involved. Hence, interest has also centered on risk taking propensity as an identifying characteristic of entrepreneurs.[36] Paradoxically, Brockhaus' study was not able, however, to identify a significantly different level of risk taking propensity between samples of entrepreneurs and managers. Nevertheless, results of a single study would not suffice to discard the hypothesis of a specific strong risk taking propensity of 'entrepreneurs' as a defining trait because risk taking is certainly a defining function.

Baumol[37] argues that entrepreneurial activity has eluded economic analysis (and clear cut definition) because, by its very (flexible, non-routine, and innovative) nature, it cannot be standardised. If it were possible to analyse an entrepreneurial activity in a general fashion, it would deprive the activity of its entrepreneurial character. Baumol cites allegorically the Heisenberg uncertainty principle: 'To observe the subject is to make it disappear.'[38] Nevertheless, while Baumol doubts the feasibility of analysis of entrepreneur decision making, he does recognise the importance of studying influences which might tend to encourage or discourage entrepreneurship.

Leadership is certainly a primary entrepreneurial function. For the U.S. Army, successful leadership implies:[39]
- knowing subordinates and showing consideration for them,
- keeping channels of communication open,
- accepting personal responsibility and setting an example for others,
- initiating and directing action,
- training men as a team, and
- making decisions.

Obviously, as we indicated above, the basic objective in entrepreneurial analysis has been and is to cultivate and nurture this kind of individual in the interest of economic development. In subsequent chapters, we shall go into depth on trait and functional analysis of entrepreneurship, making reference to concepts of Schumpeter, Knight, McClelland, Rotter, Hornaday, Chell, Schein, Cole, Kluckhohn, Shapiro, and others. As already indicated above, it will become evident that there are also numerous groups of managers and business executives who must also possess many traits and skills essential to the entrepreneur but who, in contrast to the entrepreneur, are basically bureaucratic, organisational men assuming only a moderate amount of risk or none at all.[40]

As Lewin[41] so adeptly elaborated, economists often have a paradoxical attitude toward psychological and sociological components of economic activity. As she notes, mainstream economists are very attached to the notion of rational choice, which is nonsensical if the ratio is not related to motive - a phenomenon which obviously cannot be satisfactorily explained by postulating pure utility maximisation. Motives are closely related to personal traits and functions. In order to conform to criteria for 'hard science', behaviourist schools of mainstream economy make no attempt to understand motives, claiming that such investigations do not produce 'objective' (hard) data and thus cannot contribute to better understanding reality. Only 'revealed preferences' are held by behaviourists to be, in a strict sense, empirically observable and measurable. But it is intuitively all too obvious that knowing more about socio-cultural influences on motives to start a venture, to innovate, or to adopt innovations is of the utmost importance for inciting regional economic development. Thus, the individual's motives for innovation and enterprise and patterns of functional behaviour will be addressed in subsequent chapters.

Recognising Effects of Social Structures

In all major concepts of social structures, e.g., those of Parsons, Burt, Gershuny gainful activities of the population figure among the most significant defining variables.[42] And, in turn, all economic action is embedded in structures of social relations. Quoting Granovetter:[43] 'In classical and neo-classical economicsthe fact that actors may have social relations with one another has been treated, if at all, as a frictional drag that impedes competitive markets.' Hence, in (neo-) classical economic models we find 'spot-markets' populated by rational, more or less atomised individuals in pursuit of self-interests. Although the 'homo oeconomicus' postulate is known to be untenable,[44] standard economic analysis neglects a priori identities of partners in individualised transactions of the past and present. Granovetter (and others)[45] take issue with these unrealistic propositions, illustrating continued significance of 'embeddedness' of economic transactions in modern social structures. 'The widespread preference for transacting with individuals of known reputation implies that few are actually content to rely on either generalised morality *or* institutional arrangements to guard against trouble.' [46] Continuation of one's own past dealings is valued so highly according to Granovetter because

- mutual trust based on embedded, multiplex relationships within given social structures helps reduce substantially 'transaction costs', i.e., costs involved in complex negotiations, security measures, litigation
- one trusts one's own information most - it is richer, more detailed, and known to be accurate,
- motivation of partners to be trustworthy is greater in continued relations, so as not to discourage future transactions, and
- overlying social contents of relationships carry expectations beyond bilateral relationships of trust and abstention from opportunism.

Even superficial observations of buying and selling transactions disclose that interactions rarely approximate the 'spot-market' model in classical economic theory. In fact, it usually takes a real 'jolt' to change these relationships, resulting in higher transaction costs and more risks.

The significance of embeddedness is also recognisable in the fact that many firms have interlocking directorates and their directors have multiple, closely knit business (and often private) relationships. According to Granovetter,[47] disputes between firms are frequently and preferably settled without reference to contracts or potential legal sanctions. But even public and firm bureaucracies, which ideally guarantee that fixed relationships between positions in an organisation not be influenced in their operations by incumbents, are very susceptible to personal influence. In particular, in large hierarchical organisations, (e.g., firms with long job tenures and with dense and stable networks of relationships) shared understandings and political coalitions evolve that transform the working of the apparatus.[48]

Mingione[49] differentiates between relationships of 'reciprocity' and 'association' in explaining nonopportunistic give and take and power relations between individuals within given social structures. Whereas reciprocal behaviour of individuals is defined by Mingione to be inward orientated with respect to their social group (e.g., sacrifice of individual interests to the benefit of other group members) and are instrumental in maintaining or strengthening one's own status in the group, actions based on 'association' are outward directed and aim at consolidating, advancing, or defending interests of all group members against non-members. Obviously, these relationships are of prime importance for potential entrepreneurs seeking entry to new markets, resources, etc., and are exemplified by the 'ethnic entrepreneurship' phenomenon. To cite Oliver:[50] 'Information in markets is not simply imperfect or incomplete; it is socially constructed.' Information flow and social networks are clearly crucial to business formation and survival.[51] Hence, social structures are important sources of either friction or lubrication in innovative behaviour and enterprise and merit our closer attention in subsequent chapters.

Recognising Effects of Culture

Quoting Baumol: 'One cannot reject the hypothesis that religious beliefs and cultural conditions make a substantial difference in the degree to which entrepreneurial activity is considered a desirable or undesirable way to spend one's time.' [52] Entrepreneurship is rewarded differently in different societies; profits are only one form of reward, there are various other forms of official and unofficial recognition as well. Obviously, the same is true for constraints on entrepreneurship. It is clear that rewards to and disincentives from innovation and enterprise must be perceived and subsequently evaluated by individuals according to their normative references in order to become effective.[53] At the same time, individuals differ greatly from one another and are embedded in and to a large extent are (variable) products of their society and culture. Now quoting Benedict:[54]

> No man ever looks at the world with pristine eyes. He sees it edited by a definite set of customs and institutions and ways of thinking. The life history of the individual is first and foremost an accommodation to the patterns and standards traditionally handed down in his community.

And further: 'A culture...is a more or less consistent pattern of thought and actions.' [55] Benedict directed her attention to what she calls 'cultural configurations' affecting behaviour in a constant and particular way, leading to the evolvement of basic features of modal personalities.

According to Lerner,[56] mobile (flexible) personalities are engendered by social environments characterised by a high degree of geographical mobility (yielding direct experience with/contacts to other, more advanced societies) and familiarisation with new communication media (rendering mediated experience). Communication is a defining element in culture and new media help spread new desires and learning, one of the most powerful driving forces of modernisation and enterprise. Lerner claims that Westerners have become habituated to change due to their geographic mobility and media sophistication. In fact, in the Western cultural value system change is considered to be the normal state of affairs and protection of opportunities for the individual under constantly changing circumstances to be of high priority. We shall illustrate this later using time use as an example.

Since innovations trigger technical and social change, their diffusion is of great interest to many areas of social science. Katz, Levin, and Hamilton[57] define 'diffusion' as '(1) acceptance, (2) over time, (3) of some specific item - an idea or practice, (4) by individuals, groups or other adopting units, linked to (5) specific channels of communication, (6) to a social structure, and (7) to a given systems of values, or culture.' Building on this concept of Katz, Levin, and Hamilton, we define *'diffusion' as the temporal itinerary of acceptance of some idea or item along specific channels of communication by social actors within a given social structure and cultural value system.* At the receiving end, the diffusion process is differentiated by Rogers and Shoemaker[58] according to five basic, not necessarily consecutive stages: the *awareness* stage, the *interest* stage, an *evaluation* stage, a *trial* stage, and ultimately an *adoption/non-adoption* or revaluation stage. Cultural influence on each of these stages is obvious. Rejection of an innovation might be only a mental process, but adoption is certainly not.[59] It entails a chain of also culturally governed actions to gain access to resources (manpower, premises, information, tools, material), to manage production and to market the results.

Our review of contextual factors influencing enterprise and innovation must thus examine effects of cultural configurations on the development of mobile, innovative, and enterprising modal personalities and on the functioning of diffusion processes.

Notes

[1]   Hagen 1962 as well as Riesman and others.
[2]   See e.g. Leibenstein's concept of 'molecular economics' in Leibenstein 1976, p. 3ff.
[3]   See e.g. Burt's concept of 'normative action' in Burt 1982, p. 331ff.
[4]   Lerner 1958.
[5]   McClelland 1962.
[6]   Aldrich 1979, p. 56ff.
[7]   Aldrich 1979, p. 23ff.
[8]   Aldrich 1979, p. 161ff.
[9]   Reynolds in Sexton and Kasarda 1992, p. 290ff.
[10]   Cf. Gould and Törnquist 1971, p. 148; and Sweeney 1987, p. 5ff..
[11]   Barnett 1953, p. 181.
[12]   Stöhr in Aydalot 1986, p. 31.
[13]   Hagen 1962.
[14]   Rogers and Shoemaker 1971, p.27 and 180ff.
[15]   Cf. e.g. Brown, Malecki, and Spector 1976.
[16]   Tarde 1895; Hägerstrand various works.
[17]   Arthur 1991 and 1993; Dalle et Foray 1995, p. 73; Katz and Shapiro 1985.
[18]   Camagni 1985, p. 84f.

[19] Cf. Dalle et Foray1995, p. 73.
[20] Cochran 1965.
[21] Cf. Gartner 1985.
[22] Gartner 1985 and 1989.
[23] Low and MacMillan 1988.
[24] See, e.g. Kierulff 1975, p. 40.
[25] Cf. Casson 1991, p. 22.
[26] in his Traité d'Economie Politique. Paris 1819.
[27] See observations by Kierulff 1975, p. 39; by Chell in Curran et al. 1986, p. 106ff; by Low and MacMillan 1988.
[28] Mischel 1968, p. 5ff.
[29] Mischel 1968, p. 20.
[30] Vesper 1983.
[31] Cf. e.g. Ronstadt 1984.
[32] Gartner 1989.
[33] Jenks 1949.
[34] Brockhaus 1980.
[35] De Wit 1973, p. 368.
[36] Cf. Atkinson 1957; Kogan and Wallach 1974.
[37] Baumol 1982.
[38] Baumol 1982, p.30.
[39] Cf. Gibb in Lindzey and Aronson 1969, p. 228.
[40] Cf. Collins et al. 1964, especially Chap 1.
[41] Lewin 1996. Of course, economic behaviour, on one hand, and socio-cultural and psychological phenomena interact dynamically. Nevertheless, it is certainly harnessing the cart in front of the horse proposing, as Bowles (1998) did, that the market and other economic institutions determine values, tastes and personalities.
[42] Cf. e.g. Burt 1982, p. 333; Gershuny 1983 and 1996; and Castel 1996.
[43] Granovetter 1985, p. 484.
[44] Cf. Cecora in Cecora 1994, p. 1ff and Mingione 1991, p. xxiii and 8. According to Polanyi (1944/75 and 1977; cf. also Mingione 1991, p. 3f and 22ff) pure 'market-competitive behaviour' is only one of three basic forms of exchange, the second and third forms being 'reciprocity' (e.g. among members of social networks) and 'redistribution' (e.g. by the state). Obviously, economic agents are aware of potentialities of these other forms of exchange which cannot fail to influence their market behaviour. Uncontrolled competition and opportunism on markets would destroy structures of social integration and support, thus endangering the very survival of society as a whole.
[45] Granovetter 1985; Flora et al. 1997; Johannisson et al. 1994; and Oliver 1966, p. 164ff. See also criticism of the 'individual-society dualism' in social and economic research in Mingione 1991, p. xxiv f.
[46] Granovetter 1985, p. 490.
[47] Granovetter 1985, p. 495ff.
[48] Cf. Granovetter 1985, p. 501.
[49] Mingione 1991, p. 24ff. Note the similarity to Tönnies' Gemeinschaft and Gesellschaft.

[50] Oliver 1966, p. 170.
[51] Malecki and Tootle 1996, p. 44f.
[52] Baumol 1982, p. 31.
[53] The significance of cognitive orientations in entrepreneurial behaviour, e.g. of beliefs, values, and attitudes, is discussed by, among others, Gasse in Sexton and Smilor 1986, p. 50 and Spilling 1991.
[54] Benedict 1934, p. 2.
[55] Benedict 1934, p. 42.
[56] Lerner 1958.
[57] Katz, Levin, and Hamilton 1963, p. 237.
[58] Rogers and Shoemaker 1971, p. 100f.
[59] Hägerstrand 1974, p. 19.

4 Individual Actors:

Personal Factors in Innovative Behaviour and Enterprise

The preceding chapter outlined an agenda for examining dynamics in innovative behaviour and enterprise at the grassroot level as a function of the protagonist's personality and of socio-cultural configurations of the respective environment. This chapter proposes to review in more detail personality factors (cf. sections 3.1 to 3.3), namely aspects pertaining to characteristics, socialisation, roles, and identity of the innovator or entrepreneur him- or herself. We shall be concerned with the innovative and enterprising individual, with his or her personal dispositions and with agents, values, roles, and identities contributing directly to development of the individual innovative and entrepreneurial personality. Effects of 'community', i.e., of the social and cultural setting on personality development in general, will be reserved for the ensuing chapter on social embeddedness of individuals.

Personality: Traits, Functions, Motives, and Dispositions to Act of 'Entrepreneurs', 'Innovators', and 'Adopters'

Cultural anthropologists have long been interested in the entrepreneurial personality, i.e., in discovering constitutive elements thereof in the individual and determining whether these elements or traits transcend boundaries of society and culture. Kluckhohn and Murray[1] classify determinants of personality as
- personal constitution (hereditary - biologically determined),
- idiosyncrasies (accidental - due to life events of the individual),
- personal and social role status (socially ascribed), as well as
- general community (socio-cultural - arising out of membership in some cultural group/society) affecting opportunities for and barriers to communication and mobility, 'modal' character, and value systems.

In the line of 'trait research' of the entrepreneurial personality, we find in the copious literature an ascription of certain cognitive properties deemed indispensable for the successful functioning as an entrepreneur, such as:[2] *individualism, ability to process and integrate large amounts of differing types of information, ability to choose and handle personnel, comprehension of the social and economic environment, flexibility, creativity, experience in the field* (e.g., for replication or modification of ideas from former places of work), *ability to establish and maintain contacts, good self-knowledge, good imagination, practical know-how, analytical ability, search skills, foresight, computational skills, communication skills, delegation skills,* and *organisational skills.*

With respect to motives and dispositions, two of the most frequently cited traits of entrepreneurs are *high 'achievement motivation'* and *internal 'locus of control',*[3] i.e., the subject's behaviour is assumed to be highly self-motivated. Both indicators imply that the entrepreneurial individual feels in control of his fate or of the outcome of his or her behaviour. Empirical investigations have found these characteristics to be significantly related (positively) to the level of entrepreneurial activity.[4] Using Rotter's 'internal/external locus of control scale', Shapero also found his entrepreneur sample to have a very strong internal locus of control.

Hornaday and Bunker[5] let entrepreneurs speak for themselves. Respondents were asked to rate themselves with respect to 21 characteristics on a five-point scale (five placing him or her in the upper 20% of the adult population) . A median self-rating of five was found for intelligence; creativity; level of energy; need for achievement; taking initiative, and self-reliance. A median level of four was found for risk taking; innovation; leadership effectiveness; desire for money; ability to relate effectively to others, and accuracy in perceiving reality.

Kihlstrom and Laffont proposed a 'general equilibrium entrepreneurial theory' in which entrepreneurs are modelled as *less risk averse* individuals.[6] This certainly does not mean that they are willing to take risks blindly, but entrepreneurs must be prepared to act quickly, perhaps instinctively, without exhaustively analysing situations or thoroughly planning actions because by the time opportunities are investigated they may no longer exist or may have already been utilised by competitors.[7] Knight's differentiation between risk and uncertainty is essential in this respect. For Knight[8] the term 'risk' referred to events subject to a given probability distribution (for which, in principle, insurance markets exist) and 'uncertainty' pertained to events for which the specification of probabilities is not possible. Shapero[9] took a closer look at personalities of entrepreneurs who he defined as individuals having launched a new business venture. He also found in his sample a comparatively strong willingness to accept risk. Simon[10] stressed the total self-identification of the SME-entrepreneur with the firm, and his or her vitality, perseverance, goal-orientation, and an ability to persuade, activate, and even enrapture others.

An unusually large proportion of the studied entrepreneurs were labelled by Shapero as being *displaced persons/misfits* [11] who were pushed into taking an initiative by unpleasant circumstances. Stanworth and Curran[12] cite empirical results from the USA also supporting this assumption. Schein[13] too indirectly supports this finding through his assertion that well functioning interpersonal communications among social actors tend to maintain stable social relationships, roles, and identities and, by consequence, are one of the most potent forces *against* change, such as the creation of a new venture or innovation. Interestingly, Dubini[14] found that 'negative-driven' entrepreneurs are more likely to be located in 'sparse' (i.e., disfavourable) environments, whereas 'self-actualisers' tend to concentrate in 'munificent' environments. Kets de Vries, in elaborating characteristics he had found frequently recurring in entrepreneur personalities, such as an *innate need for control*, a *persisting sense of distrust*, and a *desire for applause*, goes so far as to speak of 'the dark side of entrepreneurship'.[15]

Another frequently cited personal characteristic of entrepreneurs is *tolerance of ambiguity*, i.e., the ability to cope with uncertain situations without perceiving them as a source of threat. However, Dollinger's empirical application of a scale developed by Budner did not disclose a statistically identifiable relationship between tolerance of ambiguity and entrepreneurship.[16]

Brockhaus and Horwitz,[17] who also reviewed literature on the subject, added that entrepreneurial personalities tended to be primarily *concerned with the here-and-now*, as related, of course, to immediate problems and operations of their business. Lerner[18] sees the most adaptive persons in modern society as being those who show intellectual and affective mobility, i.e., the ability to *identify themselves with changes*, and who *strongly incline to rationality*.

Of non-personality related variables of the individual, education and experience are the two most frequently discussed. Empirical investigations have revealed experience in the field to be a major characteristic of initiators of new ventures, as discussed heretofore. On the other hand, Van Praag and Van Ophem were not able in their study to find a relationship between level of education and propensity to entrepreneurship. Differentiating between opportunities and willingness, their data analysis indicated that there be many more individuals willing to become entrepreneurs than have opportunities to do so and that while opportunities for entrepreneurship be an increasing function of age, inversely, willingness to assume this status be a decreasing function of age.[19] Obviously, the would-be-entrepreneur does not fit well into concepts of 'occupational trajectories' but is strongly influenced by the forces of social stratification.[20]

In attempting to distinguish entrepreneurs from non-entrepreneurs, trait analysis can only be of use if it can be ascertained with a high degree of statistical significance that specific traits are characteristic for entrepreneurs only and that the traits in question allow a prediction (based on probability) of entrepreneurial behaviour. We have already pointed out that many traits considered to be essential to the enterprising, innovative and independent personality are also found to be strong in other socio-professional groups, such as corporate managers. Gartner[21] points out that traits seldom have an identical or equally strong influence on social actors across all situations. Hence, the overwhelming importance of studying environmental/situational contexts of behaviour to understand the impact of possibly latent traits. This perspective corresponds closely to Mischel's findings that individuals do not demonstrate consistencies in behaviour across time and in diverse situations and thus that traits cannot be the sole base of reliable predictions on whether a person will act in a particular way in a particular situation.[22] Mischel was supported by Bem and Allen[23] who disputed, as had Allport[24] before them, that apparently differing reactions under different circumstances be indicative of inconsistencies in personal traits or personal dispositions. According to these authors, the individual controls his or her dispositions and regulates his or her behaviour according to given circumstances. Thus, for instance, in a circle of friends and acquaintances a person might well be rather extrovert in behaviour whereas as a teacher in the classroom he or she might be aloof and very formal. In this context Ajzen's assumption[25] seems rather appropriate, namely:

> that any single sample of behaviour reflects not only the influence of a relevant general disposition, but also the influence of various other factors unique to the particular occasion, situation, and action being observed.

Chell describes models assuming that individual entrepreneurial personalities change in the course of their lifetime as a function of their transactions within specific social contexts with specific reference groups and social partners.[26] Thus, understandably, Bem and Allen's conclusion on predictability[27] resembles Abraham Lincoln's comment on fooling people: you can predict some of the people some of the time but not all of the people all of the time. However, they contend that when the predictor has knowledge of the situation and of social actors' personal traits, his predictions have a higher likelihood of being accurate. Another great source of error in prediction is the effect of bounded rationality of protagonists.[28] Conlisk and Simon[29] believe that normative environments may well exert a specific bias on rationality boundedness. *Summa summarum*, these findings all indicate the significance of personality traits of potential innovators and entrepreneurs - whose potential must, however, be triggered or 'liberated' by contextual circumstances.

In analysing functions of entrepreneurs, market and social research has come up with a wide spectrum of defining activities, which differ or change in importance according to the type of venture or the current state of technology. Cole[30] specifies, for instance: *determination of business objectives, development of organisation structures and relationships, procurement of financial resources, acquisition of technical means of production, development of product market,* and *maintenance of good relations with the institutional environment.* Schumpeter[31] saw the principal role of the entrepreneur in *creative destruction* of the existing economic fabric, that is as an *innovator*; someone *identifying new strategies, products, production methods, markets, sources of supply and organisational structures and implementing these changes in his or her venture.* On the other hand, Knight[32] emphasised that the entrepreneur's major function was to *make decisions under uncertainty.* For Kirzner; Jenks, and von Mises[33] the essence of entrepreneurship was *perceiving and taking advantage of profit opportunities.*

As stated, functions of entrepreneurship may vary with the type of gainful activity but also with altering technological and social-economic contexts in the course of time. Empirical studies on entrepreneurs may restrict their samples to 'owner-managers', although managerial groups without personal ownership in corporations take what can be described as entrepreneur-like decisions.[34] Some authors do extend the term 'entrepreneur' to other groups assuming many similar functions, such as executives in corporations or managers. For instance, Lachman[35] distinguishes 'independent' from 'administrative' (managing) entrepreneurs by the degree and point at which they break away from an existing bureaucratic structure. Sexton and Bowman[36], and Schein[37] go into depth on differences in orientation between 'founder/owners' and 'professional managers/executives'. We have stated our position in Chapter 2.3 that our use of the term 'SME-entrepreneur' pertains to the owner/manager of an small or medium-sized enterprise as the prime agent of sustainable regional development at the grassroots.

Postulating that awareness or knowledge of innovations is almost always a precondition for adoption, Rogers and Shoemaker[38] imply that 'early knowers' are strategically well-placed in information flow networks and have a higher than average propensity to be 'early innovators' (they mean 'adopters'). Accordingly, they cite a number of findings on general characteristics of 'early knowers' relative to 'late knowers': They are found to have more education, a higher social status, and a higher degree of upward mobility. They are more exposed to mass media, have a greater exposure to interpersonal channels of communication, more contact to professional 'change agents', and are more likely to have a commercial orientation, as opposed to subsistence orientation. They participate more in social activity, are more cosmopolitan, and are less likely to discontinue innovations. Their economic activities are more specialised and their attitudes towards taking credit more favourable than those of their late-knowing counterparts. Directing attention back to psychological dispositions or traits of 'early innovators' (adopters) as compared to 'late innovators' (adopters), the authors cite the following personality traits:[39]

- they exhibit more empathy (ability to project themselves into other persons' situations),
- they are less dogmatic,
- they exhibit more rationality,
- they are more intelligent,
- they are more inclined to accept change,
- they are less averse to risk,
- they are more favourably inclined to education and science,
- they exhibit higher levels of achievement motivation,
- they are less fatalistic (internal locus-of-control), and
- they have higher levels of aspiration.

The Family Household and Socialisation of Individuals

In general, even economists recognise that there are (mostly unmeasured) family factors influencing the level of education and occupational status of individuals.[40] And, of course, in all sociological, psychological, and anthropological schools of thought the family is attributed a major role in personality formation and socialisation of the individual.[41] De Vos and Hippler[42] state that socialisation is usually the central focal point of cross-cultural research on personality. Kardiner, Benedict, Inkeles and Levinson, Cochran and others have defined the family as a primary institution conditioning early experiences to produce culturally determined basic personality types, and occasional deviations from these specific patterns. But, as we shall see in a subsequent discussion of actors' embeddedness in social configurations, individual variations and deviations from prevailing personality types do not disprove existence of 'modal personalities' specific to given cultures.[43]

The importance of family background for formation of the 'entre-preneurial personality' was recognised by Kets de Vries,[44] who noted that early life experiences shape traits like impulsivity, persistent feelings of discontent, rejection, powerlessness, and low self-esteem, shaping later behavioural patterns. Thus, similar to the 'push' concept of Shapero,[45] he considers that frequently the entrepreneurial personality is pushed towards a more attractive and socially esteemed role and identity than was accorded him or her by the family environment. Over half of entrepreneur-respon-dents to an investigation by Hornaday and Bunker[46] referred to depri-vation in their childhood as a factor in their strong desire for compensation of some kind. On a more positive strain, McClelland cites a number of studies indicating that parents' high achievement standards and encour-agement, warmth in the home, and a nondominating, nonauthoritarian fa-ther be three important factors for developing a 'need for achievement' in boys.[47]

Shapero and Sokol[48] state that social and cultural value systems are significant factors leading to 'entrepreneurial events' (ventures) and thus also underline in this context the importance of the family for transmitting values to its members and offering them 'credible' role models for enter-prise. There appear to be identical patterns across cultures: A majority of entrepreneurs have parents, in most cases fathers, who were either self-employed or entrepreneurs with employees. Shapero[49] corroborated num-erous other findings[50] showing that *family role models* (e.g., father had his own business) and *personal acquaintance* (with others having started their own companies) were a forceful factor in inclination to entrepreneurship. Cooper,[51] who distinguished between entrepreneurs in 'high-tech' and 'non-technical' businesses in his analyses, found role models to be important for both groups.

Another key role is attributed by Shapero and Sokol, Aldrich and Zimmer, Cooper and Dunkelberg, Hansen, and Pennings[52] to families and kin as providers of financial support for ventures, a phenomenon which works particularly well in families from ethnic groups due to both highly 'dense' internal network links between members and the salience of their group's identity and contours with respect to the host community. Data show that the great majority of company set-ups use capital from personal savings and borrowings, particularly from relatives. This 'informal' finan-cial market greatly reduces 'transaction costs'[53] but requires a high degree of trust and family coherence, which is often more frequently found in certain ethnic minorities or immigrant groups. Even when connections are sought outside the family, informal transactions, i.e., with other individuals from the local community, are the major source of financial support. Informal organisation of financial capital for ventures is only part of the enormous entrepreneurial potential inherent to an individual social actor's family. In a socio-economic environment characterised by 'over-regulation', family relationships based on mutual trust, high motivation, solidarity, and benefit-sharing prove to be an invaluable human asset for starting and operating a SME-business. In particular, huge 'transaction costs' involved with payroll labour (e.g., contract negotiations, labour monitoring, legal affairs, formal accounting) can be largely avoided while retaining a high degree of flexibility (e.g., work scheduling, overtime, delegation of responsibility and authority). Furthermore, through family members the extent of the entrepreneur's reliable social network can be extended considerably, as well as the reach of his or her 'information fields' and 'action spaces'. The extended family is assuredly an institution of prime importance for sustainable regional development which has been widely neglected in highly developed western societies, which have tended to cultivate dependencies of individuals for social security and gainful employment on the market or the state instead of on family solidarity.

U.S. research found that in their occupational orientation female entrepreneurs are four times more likely to have been subject to influence of an entrepreneurial parent than members of the general population.[54] However, with respect to family continuity, it was shown that less that half of female entrepreneurs were married or had an equivalent stable relationship as contrasted to more than 90% of male entrepreneurs. Conjugal relationships were more heavily strained when the woman was the entrepreneur. The role of the conjugal partner was generally a stabilising factor in male-led businesses, but not in female-led businesses.

In investigation of communication structures within family households, Komarovsky[55] noted that transition in modern Western societies from a 'traditional' to a 'companionship' form of family exhibiting joint decision-making and task-sharing was furthest advanced in middle-class house-holds. Traditional household patterns had shown stronger separation of responsibilities and weaker interpersonal communication. Considering the above discussion of personal traits of entrepreneurs, this might suggest a weakening of individual potential for enterprise in Western middle class family households. To date there has been no evidence corroborating this supposition, however.

The Individual's Role, Identity, Values and Attitudes as Behavioural Determinants

Social roles and social statuses are major building blocks in concepts of social structure. According to Parsons[56] social systems are essentially based on *differentiated roles* of their members. In interaction of social partners *patterned role expectations* govern the individual's actions.[57] These expectations are coupled with implicit, sometimes explicit rewards for conforming or retributions for deviant behaviour. In well integrated societies, these expectations have been institutionalised in *prevailing values* (norms) making subconscious imprints on the individual's personality and identity. Overall, the greater the degree of social actors' conformity to norms and of their integration, the more static a society is likely to be.[58] Hence, one of the most important seeds of social change is failure to maintain social integration of personal and social roles within the generalised value system.[59] Obviously, innovators and entrepreneurs are agents of change often assuming somewhat deviant roles. 'Creative destruction'[60] of the existing social and economic fabric cannot fail to engender conflict and resistance by vested interests in the status quo.[61] Rewards gained by the deviant social actor must, in such cases, more than compensate encountered social disincentives for role deviation. Again, some form of value orientation is inherent in all social role expectations and behaviour;[62] studies of roles and values are closely entwined. An individual's attitudes towards role expectations may vary according to whether the role is ascribed, i.e., acquired automatically and involuntarily, e.g., family membership, sex, order of birth , or achieved, i.e., as the result of a conscious effort or achievement, e.g., becoming an entrepreneur.[63] For Linton, values and attitudes are an important measuring rod for both individual and for basic, socio-cultural (modal) personalities.[64] Besides subconscious personality traits, expressed values and attitudes are major sets of variables having been found to be related to acceptance of innovations[65] and to disposition to be an entrepreneur. Since innovation and entrepreneurship imply deviation from the status quo and, to a certain extent, from supra-individual value systems[66] it cannot fail to have repercussions for both the normative framework and the individual's behaviour. Discrepancies between personal values of the innovator or entrepreneur and prevailing norms create tension between the agent and his or her environment which affect social approval of the agent's actions. In examining the innovation diffusion process, Rogers and Shoemaker[67] borrow Festinger's term *'cognitive dissonance'* to describe uneasiness arising from discrepancy between an individual's attitude towards a specific innovation and an incompatible decision to either adopt or reject it.

Jenks[68] describes the 'firm' as a microcosmos having its own culture influencing patterns of reciprocal behaviour of its members. According to Jenks[69] individuals learn a personal role as entrepreneur by perceiving and interpreting manifest expectations of social partners in business situations, by experiencing sanctions for more or less appropriate responses to social partners, by adopting role models and by adapting available resources to the unique situation at hand (learning by doing).

Classifications of the *entrepreneurial identity* often are based on either specific functions it fulfils or on personal attitudes and management styles. For example, Stanworth and Curran[70] propose: 'artisan' identity (emphasis on autonomy, personalised service, product quality), 'classical entrepreneur' identity (profit orientation), and 'manager' identity (aims at organisational optimum and fulfilment of goals). Hence an entrepreneur's overall social identity is based on characteristics of both personality and status. *Social status* involves designated rights and obligations of the individual.[71] Merton[72] points out that each status is connected with an array of roles which he calls a 'role-set'. A fundamental postulate of role theory is that a person's attitudes are subject to influence by the roles he or she occupies in a social system.[73] The importance of social status should not be underestimated. Some members of a group may do with impunity what others may not do at all. An experiment by Hollander[74] showed that behaviours reflecting innovative action were found to be disapproved significantly less the higher the status attributed to the innovator. However, Gerschenkron[75] postulates that general social approval of entrepreneurial activity can, but must not necessarily significantly affect its volume and quality. He notes that there is a certain degree of diversity and a latitude of action in any normative system. Furthermore, as the discussion of personal traits of entrepreneurs disclosed, they are perfectly capable of coping with conflict situations. On the other hand, in cases of receptive social contexts, the innovator or entrepreneur may even prove to be a welcome initiator of action in response to challenges arising in the economic and social environment.

Creating new organisations, especially innovative kinds of organisations, also creates entirely new roles and communication channels within existing organisational structures and within the general socio-economic environment.[76] Many new social relations arise between strangers; rules (e.g., rewards and sanctions), delegation of responsibilities, foundations of trust and obligation must evolve on a trial and error basis. MacMillan[77] emphasises the importance of political skills for forging keystone alliances and network building with respect to information access, anticipating and dealing with reluctance, and mobilising resources. Thus, it is obvious that an individual's specific role set and identity, as well as his or her personal values and attitudes are important determinants of whether he or she will become an innovator and/or an entrepreneur. On the other hand, it is also evident that success, particularly as an entrepreneur, depends on participation in the socio-economic and cultural environment.

Notes

[1] Kluckhohn and Murray 1953.

[2] Barnett 1953, p. 65; McGaffey and Christy 1975; Casson 1982, p. 29ff; and Bhide 1994.

[3] For *achievement motivation* see McClelland 1976. For *locus-of-control* see Rotter 1966; Van Praag and Van Ophem 1995, p. 524f.

[4] Durand and Shea 1974; Evans and Leighton 1989, p. 521; Lachman 1980.

[5] Hornaday and Bunker 1970, p. 50f.

[6] Kihlstrom and Laffont 1979.

[7] Cf. Bhide 1994.

[8] Knight 1921. See also LeRoy and Singell 1987, p. 395f; Barreto 1989, p. 39ff.

[9] Shapero 1975.

[10] Simon 1996, p. 10.

[11] As a matter of fact, some market researchers consider 'having been fired at least once, better several times' as a hallmark of 'entrepreneurs'. At the very least, frequency of former job changes has a positive effect on willingness to become an entrepreneur - see e.g. Van Praag and Van Ophem 1995, p. 528.

[12] Stanworth and Curran 1976, p. 102.

[13] Schein 1960.

[14] Dubini 1988.

[15] Kets de Vries 1985 and 1986.

[16] Dollinger 1983.

[17] Brockhaus and Horwitz in Sexton and Smilor 1986, p. 33.

[18] Lerner 1958.

[19] Van Praag and Van Ophem 1995, p. 528 and 530.

[20] Cf. e.g. Gershuny 1996.

[21] Gartner 1989, p. 31.

[22] Cf. Mischel 1973.

[23] Bem and Allen 1974.

[24] Allport 1966 and 1937.

[25] Ajzen 1991, p. 180.

[26] Chell in Curran et al. 1986, p. 106ff.

[27] Bem and Allen 1974, p. 517.

[28] Cf. Simon 1969 and 1982; and more recently Conlisk 1996.

[29] Conlisk 1996, p. 676f and Simon 1993.

[30] Cole 1967, p. 35.

[31] Schumpeter 1949; Cf. also Barreto 1989, p. 5 and 22ff.

[32] Knight 1921; and LeRoy and Singell 1987. Cf. also Barreto 1989, p. 5.

33   Kirzner 1974; Kirzner in Kent, Sexton and Vesper 1982; Jenks 1949, p. 128; Von
      Mises 1949. Cf. also Barreto 1989. p. 15 & 17.
34   Cf. Jenks 1949, p. 110f.
35   Lachman 1980, p. 109.
36   Sexton and Bowman 1985.
37   Schein 1983, p. 26.
38   Rogers and Shoemaker 1971, p. 107ff, 185ff. See also Kegerreis, Engel, and Black-
      well 1970.
39   Rogers and Shoemaker 1971, p. 187ff.
40   De Graaf and Huinink 1992.
41   Cf. Gershuny 1996, p. 177 and numerous others cited in Cécora 1993 and 1994.
42   De Vos and Hippler in Lindzey and Aronsson 1969, p. 359ff.
43   Cf. Inkeles and Levinson in Lindzey and Aronson 1969; Benedict in Kluckhohn and
      Murray 1953, p. 522ff; Gorer in Kluckhohn and Murray 1953, p. 246ff; and Inkeles
      in Kluckhohn and Murray 1953, p. 577ff.
44   Kets de Vries 1985; 1986; Cf. also Chell in Curran et al. 1986.
45   Which will be elaborated in discussing effects of social configurations in a subsequent
      chapter.
46   Hornaday and Bunker 1970, p.51.
47   McClelland 1976 and 1962, p. 110. See also Hornaday and Aboud 1971, p. 147.
48   Shapero and Sokol in Kent, Sexton, and Vesper 1982, p. 83ff.
49   Shapero 1975.
50   For instance, Van Praag and Van Ophem 1995, p. 522; Brockhaus in Kent, Sexton,
      and Vesper 1982, p. 52; Kourilsky 1980.
51   Cooper in Sexton and Smilor 1986.
52   Shapero and Sokol in Kent, Sexton, and Vesper 1982, p. 86; Aldrich and Zimmer in
      Sexton and Smilor 1986, p. 14f; Cooper and Dunkelberg 1987, p.14f; Hansen 1992,
      p.99; Pennings in Kimberly, Miles et al. 1981, p. 156.
53   Cf. Williamson 1981 and Pollak 1985.
54   Watkins and Watkins in Curran et al. 1986. Furthermore, findings showed that
      female entrepreneurs benefited much less than males from the general educational
      system with respect to knowledge and skills relevant to their venture. It was their
      professional work experience and qualifications obtained after termination of formal
      education which proved to be much more important.
55   Komarovsky 1961.
56   Cf. Parsons 1951.
57   Cf. Merton 1957, p. 110.
58   Cf. Gerschenkron 1953, p. 3.
59   Parsons 1951, p. 179 and Jenks 1949.
60   Schumpeter 1942.
61   Cf. Gatewood, Hoy, and Spindler in Hornaday, Shild, Timmons, and Vesper 1984, p.
      265ff.
62   Cf. De Vos and Hippler in Lindzey and Aronson 1969, p. 366.
63   Cf. Alexander 1967.
64   Linton 1981. Cf. also Inkeles & Levinson in Lindzey and Aronson 1969, p. 430.
65   Katz, Levin, and Hamilton 1963, p. 249.
66   Schumpeter 1926, p. 134.

67   Rogers and Shoemaker 1971, p. 112.
68   Jenks 1949, p. 114f.
69   Jenks 1949, p. 138f.
70   Stanworth and Curran 1976, p. 104.
71   Cf. Merton 1957, p. 110.
72   Merton 1957, p. 110f.
73   Cf. Lieberman 1956.
74   Hollander 1961.
75   Gerschenkron 1953, p. 6 and 13f.
76   Stinchcombe in March 1965, p. 148f.
77   MacMillan 1983.

5 Contexts:

Socio-Cultural Embeddedness and Its Implications
for, Opportunities for and Constraints on
Innovative Behaviour and Enterprise

Having discussed in sections 3.4 and 3.5 and Chapter 4 the importance of
personal factors for the individual's predisposition to innovative and entrepreneurial activity, we now devote closer attention to significance of social
structures and cultural configurations identified in sections 3.4 and 3.5 as
having strong effects on innovative and entrepreneurial behaviour.

Cartwright[1] observed that, by the end of the 1950s, the focus of most
empirical research on entrepreneurship had shifted from personal
characteristics of individuals to situational influences in interaction with the
individual's 'position', 'function', or 'role' in society. According to
Aldrich and Zimmer,[2] the causes of this shift away from personality-based
approaches are:
- failure to identify a truly generic 'leadership' trait,
- failure to identify 'leaders' outside of the context where leadership is
  self-evident,
- overstatement of the extent to which entrepreneurs differ in respect to
  specific traits from other socio-occupational groups, and
- apparent underestimation of actually available entrepreneurship potential by past studies.

These shortcomings of research concentrating solely on the entrepreneurial
personality are evident. Nevertheless, as we have explained above, they do
not prove irrelevance of personal traits for investigations of innovation,
enterprise and regional development but rather highlight significance of
taking into account the social actor's embeddedness in a socio-economic
and cultural environment.

In 'days gone by', regional development potential was seen to be
primarily a function of diversification in structure of regional economic
activities and of qualifications of the regional working force. Of course,
monostructured regional economies may be very vulnerable to specific sets
of unfavourable circumstances but in general, although indisputably very
significant, economic structures are now viewed as being less important for
future prospects of the region than generally prevailing attitudes in the
region towards work and learning as values, toward enterprise, innovation
and change, and towards personal initiative, cooperation and partnership,
on one hand, and characteristics and dynamics of communities and social

networks, on the other.[3] For example, according to Nittykangas,[4] the rural milieu (in Finland) still has an effect on personal traits and values of local populations contrasting with those of urban populations. In particular, the author contends that their locus-of-control is more external than in urban groups, partly because the agricultural sector, in which many are gainfully active, depends on factors outside of regional control. Although the farming population has basically a positive attitude toward enterprise and feels as part of the business culture, its behaviour is by consequence of income dependency on administrative decisions not market-oriented and suspicion of new ideas is considered to be a typical characteristic of rural Finns in general. Similar patterns might be observed in other countries, too, although there are not many developed countries with such strongly isolated areas.

In identifying factors permitting prediction of new firm births, Reynolds'[5] empirical studies gave strong testimony to potency of regional factors in influencing regional firm birth rates by stressing the importance of

- socio-structural variables (size and composition of populations, including age, gender, ethnic origin)
- norms of society and culture (tastes and interests, cultural contexts)
- institutions and power structures (legal and political contexts) and
- social networks and peer groups (informal affiliations).

Hence situational influences on innovation, enterprise, and consequently on regional development result from regional sets of opportunities and constraints emanating from these factors. Nevertheless, opportunities are irrelevant for economic development unless recognised and utilised and our review of empirical findings has already shown that people obviously vary widely in their ability to recognise and to seize opportunities or to deal with constraints. Research has indicated that entrepreneurs are more active than others and tend to display 'opportunity-seeking' behaviour.[6] The following chapters spell out the importance for entrepreneurship and innovation of socio-economic and cultural contexts in the recognition and utilisation of opportunities. These factors are structural features of the social environment, cultural configurations, local and regional power structures, the individual's social networks and peer groups, and resultant patterns of communication and information flow.

Socio-Structural Features of 'Innovative Environments'

Noting that entrepreneurs do not operate in socio-economic vacuums, Gartner[7] reviewed literature examining 'pushes' and 'pulls' exerted by environments on organisations and their initiators.[8] In organisation theory literature he found two basic approaches to analysing environmental effects: The first, *environmental determinism*, envisions the environment as a set of external factors to which the organisation or entrepreneur must adapt to survive. This would seem to apply only to already existing and geographically immobile enterprises. The second approach, called *strategic choice* of organisations, would seem to be more in line with today's (internationally) mobile capital; it assumes a greater potential for action and self-determination by existent organisations able to seek out and identify environmental factors best suited to their needs, in other words: it assumes that siting decisions are taken by entrepreneurs weighing (rationally and objectively) contextual opportunities and constraints.

Bruno and Tyebee[9] reviewed literature identifying *environmental factors stimulating new-born entrepreneurship* and enumerated as important factors, aside from material/infrastructural and political prerequisites for the creation of new ventures, the regional population's positive attitudes towards entrepreneurship, and its relevant knowledge, skills, and experience. The authors demonstrated the significance of these factors by pointing to exemplary cases of entrepreneurial action, many being spin-offs from near-by incubator organisations. Empirical investigation by Kilkenny et al.[10] showed that 'reciprocal community support' (support by community in response to business donations and services to the community) makes a significantly positive contribution to business success. The GREMI research group[11] has been elaborating new concepts of economic space in an attempt to define indicators of 'milieux innovateurs' based on territorial frameworks of relationships, personal interactions and synergies, and collective socio-economic action. They have examined attitudes, information flow, learning and imitation processes, form and intensity of personal contact and cooperation specific to local environments.

Obviously, there are synergies between innovation and enterprise on one hand and socio-structural features of locations on the other. Innovation leads to socio-economic changes modifying society's structures, values and aspirations of the population and resetting the stage for further innovation adoption processes.[12] According to Rogers and Shoemaker,[13] social structure can impede or facilitate the rate of innovation diffusion by so-called 'system effects'. General norms, status-related values and identities, social roles and hierarchies represent system-immanent factors influencing the individual's behaviour. Ever since Triplett's study demonstrated that individuals tend to turn a crank faster when they are in the presence of co-workers and Durkheim argued that patterns of small-group interactions and face-to-face associations among peers be determining factors in human behaviour, social structure has been a focal point in social psychology.[14]

Brown et al.[15] note that two important approaches to the study of innovation diffusion are based on significance of locational social structures for rapidity in adoption of new processes and products: Communications-based approaches highlight structures in information flow while approaches focussing on market infrastructure stress actual availability to potential adopters of the innovation or of its prerequisites (facilitated e.g. through the medium of 'diffusion agencies'). Brown et al. found that availability of good information and efficiency of diffusion agencies may well be even more decisive factors for time of adoption than innovativeness-traits of potential adopters. In effect, people make decisions within and in interaction with social networks consisting of family, friends, co-workers, employers, casual acquaintances, etc.[16] It is within these social structures that information and resource flows are effectuated between individual social actors in acts of bestowing, exchange, dependency, competition, brokerage, or coalition.[17]

Katz, Levin, and Hamilton[18] pointed out that considerably more research has been done on implications of structures in social relations for diffusion of innovations across boundaries than on diffusion within boundaries. Very few studies have been concerned with implications of structural arrangements within a group situation and, in these studies, socio-structural characteristics are most frequently used to describe individuals' roles and identities within the context rather than to describe the context as a whole.. This is all the more peculiar because, as society becomes more complex, social relationships become more specialised and fragmentised.

Quoting Stinchcombe:[19] 'It is common knowledge that "modern" societies carry on much more of their life in special-purpose organisations than do "traditional" societies. Any one of the numerous ways of dividing societies into "modern" and "traditional" gives the same result: wealthier societies, more literate societies, more urban societies, societies using more energy per capita, all carry on more of their life in special-purpose organisations, while poor, or illiterate, or rural, or technically backward societies use more functionally diffuse social structures.' Stinchcombe indicates that there is some, mostly indirect empirical evidence of a positive relationship between a population's level of literacy, degree of urbanisation, level of organisational experience, and the rate of organisation formation in society. Later, we shall describe how distribution of power is an aspect of social structure determining whether individuals have access to information/new ideas, whether the individual and his or her social group benefit from the changes initiated (prestige, money, power), whether the individual can obtain the necessary resources, and whether the innovative person can defeat or form alliances with those with vested interests in the status quo.[20]

Stinchcombe also pointed out that creation of new organisations necessitates evolution of relationships between strangers. Higher levels of literacy and education facilitate the individual's social competency (with unknown persons and institutions) as do highly differentiated urban environments. Rural environments are thus ceteris paribus less amenable to the birth of new ventures. Socio-economic environments with larger numbers of small firms, i.e. with higher degrees of agglomeration and diversification, are more receptive and attractive to newcomers and have comparably low entry barriers.[21] The greater regional market shares of large firms, the higher barriers to new entries are. Urbanisation has been especially conducive to creation of new organisations due to relatively easy access to relevant resources.[22] This has been due to competitive advantages as compared to rural areas in access to information which, however, in the electronic age have declined to a certain extent. In the early 1980s, Pennings still saw population density as being perhaps the most important predictor of organisational birth-rates. Besides implied greater political influence of larger communities, Pennings attributed this fact to spatial adjacency with spill-over effects and to greater exposure to a large and heterogeneous pool of persons and organisations, often inciting nonconventional alternatives. It remains to be seen how Internet will affect in the long-run comparative advantages of non-global agglomerations.

Political hierarchies and programmes are of decisive importance for the creation and maintenance of regional structures. Political systems exert, of course, a major influence on opportunities for and constraints on entrepreneurship and innovation. Push and pull effects towards entrepreneurship were exemplified in comparative analysis by K. Roberts and B. Jung[23] of self-employment of young adults in Britain and in a Poland in the midst of political and economic transition. Forces of both push and pull were obviously greater in Poland and resulted in a clearly higher proportion there of school graduates (neither pursuing higher education nor unemployed) who were self-employed. Attitudes of Polish youth towards entrepreneurship were very positive, although many would have preferred good dependent employment if it were available. Political and economic transition made 'informal' activities easier and especially lucrative. 'Informal' activities have proven to be an effective and much used training ground for entrepreneurship.

Almond and Verba[24] differentiated prevailing political systems according to the population's general level of knowledge about society and the political system and its self-perception as social actors within the system, in particular regarding 'subjective competence' or the ability to exert political influence[25] and even made attempts at classifying certain national societies.[26] According to Shapero and Sokol,[27] company formation rates are much higher in the U.S. than in France and Italy. Lower company formation rates are probably to be found in most other European countries as well as an effect of political cultures characterised by a comparatively strong regulatory environment.

Cultural Configurations and Patterns of Behaviour

When investigation of economic behaviour or regional development takes place within a single cultural environment accepted behavioural patterns and role expectations are subconsciously taken for granted by social actors and analysts and, in most cases, adhered to or taken into account. Coping with a 'normal' scope of deviation from these patterns is commonplace and does not pose particular problems. Measures affecting populations of two or more cultures, such as trans-/supranational regional development programmes of the European Union or situations where planning experts belong to one culture and the indigenous population to another, are different matters altogether. Intuition and observations clearly indicate that culture is a determinant force in perception, as well as in all human motivation and behavioural patterns, including those we call 'economic', and hence merits our close attention. In spite of economic globalisation, forces of culture are rarely a topic outside of anthropology and ethnology and their interface with sociology. Economics appears helpless in dealing with such phenomena and thus, with few exceptions,[28] ignores them completely. For this reason, it is appropriate to begin with an extended excursion into basic cultural mechanisms and their influence on behaviour in general before addressing their effects on innovation and enterprise in particular.

Unfortunately, it is not possible to find a universally accepted definition of 'culture' which is, in any case, more comprehensive than the term 'society'. Common criteria for defining cultures are geographical boundaries, mutually intelligible languages, and sometimes basic similarities of institutions, economic systems, etc.[29] These delineation criteria are rather superficial and do not touch cultural content. Linton defines and differentiates 'society' and 'culture' as follows:[30] A 'society' *is an organised group* of people having learned to work together. A 'culture' *is a configuration of learned behaviour patterns* whose components are shared and transmitted by members of a particular society. It can be shown, however, that these patterns may differ according to the individual's roles and status in the respective social structure or to his or her adherence to individual subcultures or classes. De Vos and Hippler[31] reviewed many studies indicating cultural variations in perception and cognition. They were careful to specify that there were no absolute differences in perception and cognition between compared groups but rather significantly differential incidences in 'modal' [32] patterns.

E.T. Hall[33] described culture as a set of deep, common, unstated experiences which members of a cultural group share and communicate unconsciously and which form the backdrop against which all other events are judged. For Kroeber and Kluckhorn:[34] 'Culture consists of patterns.. of and for behaviour, acquired and transmitted by symbols constituting the distinctive achievement of human groups, including their embodiment in artefacts; the essential core of culture consists of traditional.. ideas and especially their attached values; culture systems may.. be considered as products of action .. (and) .. as conditioning elements of future action.' Important parameters of culture are thus: *shared perception, subconscious formation/transmission/communication of common values and attitudes*, specific *group structures and institutions*, similar *patterns of individual behavioural dispositions* and of reactions to stimuli.[35] Intergenerational transmission of behavioural patterns were shown in experiments by Jacobs and Campbell[36] to pass on 'significant' remnants of response patterns through four or five generations.

In 1934, a prominent anthropologist, R. Benedict, had already gone beyond description of individual behaviour as an outcome of monolithic cultural influence. Her approach was to circumscribe psychological coherence of culture and social institutions as a whole with her concept of 'cultural configurations'.[37] Objections to her concept seldom questioned its general validity, but rather criticised insufficiency of empirical evidence[38] or mistakenly considered it a strictly 'deterministic' approach to the influence of culture on individuals.[39] The cultural configuration concept proves to be flexible and applicable to a wide variety of circumstances. For this reason, we choose to use it in elaborating cultural impacts on innovation and enterprise.

Parsons and White,[40] considered values to be part of the cultural system, not of the social system. However, this exclusivity is heavily contested and hardly tenable in view of difficulties in delineating the two systems in reality and of the fact that both systems are by no means monostructural. For instance, Tajfel[41] elaborates on the socio-cultural determination of perception, values, and attitudes relative to the environment. Specific sets of stimuli may be perceived differently within differing social classes or cultural contexts engendering differences in perception, evaluation, and expectancy. These processes provide the basis for 'social stereotypes'; members of stereotyped groups are expected to be similar to one another with respect to the characteristics or behavioural dispositions in question and to differ accordingly from members of other groups. Tajfel[42] concludes that culture determines to a large extent an individual's exposure to sets of stimuli and his or her familiarity with and evaluation of them.

The most important cultural variable affecting perceptual phenomena is estimated by F. Boas, E.T. Hall, and H. Tajfel[43] to be the prevalent system of communications. Beside given *structures and media of communication*, which will be addressed in a subsequent chapter, *language* is an obvious and most important example, seconded by specific information codes in common *visual and auditory symbols, accepted modes of expression* in art, oral representation, etc. In cognitive processes, significance of language features for thought formation and transmission are obvious.[44] For example, languages impose different segmentations of sensory experience,

• in the number of labels/categories assigned to a continuum,
• in the contours and contrast between categories, and
• in the consistency with which a particular label is assigned to a particular stimulus on the continuum.

Two languages differ in programming and differentiating the same events/objects. Since language molds structures of perception and thought no belief or philosophical system should be considered disregarding language. But even language carries only part of the message: What is taken for granted and not spoken in a given culture is 'filled in' by the listener. Knowledge of culturally specific visual, auditory, and even olfactory and tactile symbols can also be essential for understanding perception (filtering/ranking/ordering) of sensory worlds and for interpreting communications.[45] Probably the most differentiated analysis of the multiple forms of culturally variable communication was offered by E.T. Hall[46] with his Primary Message Systems in Communication. He stresses the difficulty in identifying and interpreting these systems because culture tends to hide more than it reveals, and most effectively from its own members.

Hence, new items (e.g. innovations) and entrepreneurial opportunities must not only be physically but also psychologically available; culture may well determine whether their presence is perceived. Thus, in Cochran's view, [47] culture may be conceived as an important intervening variable between existing environmental factors and the forces of change, i.e. to innovation or new ventures.

Cultural Specificities in Awareness and Use of Time as an Example

Comparisons have indeed revealed enormous differences between cultural groups in perception and attitudes which are imperceptible to persons within one culture because the culturally prevalent image or attitude is considered to be 'natural' or a reflection of 'reality'. Let us exemplify this phenomenon by taking a look at mono- and polychronic time systems which prevail to differing degrees in different cultures, subcultures and other segments of populations: Since time is a valuable and limited resource of great import for innovation (e.g. innovations for saving time) and enterprise (e.g. manager's productive/efficient use of own and employees' time) it is a very appropriate example of cultural effects on economic behaviour.

Max Weber's thoughts on rationality, the 'Protestant Ethic' and the spirit of capitalism captioned the evolving state of mind and prevailing attitude towards time use in cultures of industrialising countries, especially in Northern Europe and North America, marked by a 'consciousness of time' awakened notably by Calvin and Benjamin Franklin. Attitudes on the efficient and morally acceptable use of the divine gift 'time', under-stood to be a limited resource for production, culminated in the all-pervading concept of 'productivity' of time, which is defined as work output per period of time input. Wasting time was considered a sin; increasing productivity became a virtue and an explicit socio-economic and political goal. As is well known, a major characteristic of the industrialisation process was increased division of labour with strong specialisation resulting in, as Marx put it, an 'alienation' of the worker from the fruits of his labour. This had distinct ramifications, not only for culturally specific patterns of time use, but also for culturally predominant perceptions of and attitudes toward time. In describing these ramifications, I shall elaborate and extend the concept of the renowned cultural anthropologist, E.T. Hall,[48] see Table 5.1.

Hall identifies time as a core system of all cultures, each culture having its own specific time frame. The corresponding concepts, rules and patterns of behaviour are unspoken, implicit or hidden paradigms of what he terms 'primary level culture'. In other words, they have been internalised to such an extent that they belong to subconscious structures of perception and behaviour. The prevalent time system is practically unrecognisable to anyone within his or her own cultural framework - it is when one is confronted with culturally determined perceptions and behavioural patterns in other societies, often resulting in conflicts, that the astute observer begins to discern cultural specificities of his or her own social environment. This is a situation analogous to Einstein's discovery of relativity of time being a result of the study of an extra-terrestrial phenomenon (Saturn's orbit). Furthermore, Hall[49] notes that individuals are dominated in behaviour by complex hierarchies of interlocking cycles and rhythms, some of which are related to cultural influences, others to nature and biological processes. Culture is thus an important determinant of whether two persons are in synchrony in 'the dance of life', a keystone in well working personal relationships. Hence, as Hall points out, many societies have functioned reasonably well without notions of sequences or simultaneousness of events[50] or even without having a word in their language for 'time', 'being late' or 'waiting'[51]. In the following scheme we have extended somewhat and juxtaposed Hall's concepts of mono-chronic and polychronic time systems.

Unsurprisingly, Hall classifies cultures in Britain, in English-speaking countries in North America and Oceania, and in Northwestern Europe as having essentially monochronic time systems. Cultures, of course, are not monolithic and there are indeed segments of these cultures exhibiting some traits and life styles characteristic of polychronic time systems, which is notably the case of housewives, of members of farm family households, and of agents of bureaucracy. Hall ascribed polychronic time systems to cultures in Latin America, in Turkish and Arabian countries, in India, and to some extent to Mediterranean cultures.

Obviously, monochronic time systems reflect cultural processes according to time concepts of Calvin, Franklin and Weber. We can assume that without the prevailing monochronic time system, industrial civilisation, characterised by rapid innovation cycles, expansive entrepreneurship and corporate culture, could not have evolved so quickly. Nevertheless, the 'march of progress' towards industrialised, technically highly sophisticated civilisation does not automatically imply an irreversible trend from polychronic to mono-chronic time systems. To the contrary, redundancy of ever larger parts of the work force would tend to further revival of polychronic time use patterns and certain measures taken to restructure enterprises (e.g. 'lean management' with semi-autonomous work groups) take into account advantages of polychronic systems which are more likely to facilitate recognition of individual activities as parts of a whole, assuring that each function is fulfilled and leaving a large degree of discretion on how and when activities are performed to individuals responsible. It is thus of the utmost importance to realise, with Hall[52], that monochronic time systems may well be thoroughly learned and subconsciously anchored but they are basically arbitrary and culturally imposed. In any event, they do not represent the natural and logical way to organise life, they are not inherent to man's biological rhythms or creative drives, and they are indeed subject to change.

Table 5.1 Culturally determined time systems
(author's elaboration of E.T. Hall's concept)

Monochronic time system	Polychronic time system
• Concentration on 'one thing at a time'; compartmentalisation permits only selected perception/reaction inhibiting recognition of contexts, segmentation less awareness of being part of whole	• 'Simultaneousness' as a way of life; primacy of whole contexts and social interaction in groups, involvement with people, almost never alone
• Orientation to the future, the past valued chiefly as a source of experience	• The present is most important; promise of future benefits is irrelevant to present actions
• Low appreciation for routine / repetitions (= 'ruts'); preference for surprises, change, the 'new'	• Appreciation of tradition; at home in routine + repetitious situations where time, activities, locations are syncopated and fixed
• Concept of discreteness/limitation of time	• Time is 'here and now'
• Time is similar to an object that can be 'saved', 'wasted', 'lost', 'spent'	• Intangibility of time
• Concept of time as a continuous band that can be segmented and	• Time is indivisible; scheduling very difficult

scheduled

• Strong functional structuralisation of time; strong control of activities and inter personal interaction	• Time is largely unstructured; mixing of activities and interpersonal relations
• Privacy in interpersonal relations/meetings	• Lack of privacy; communications in groups
• Importance of appointments, schedules, promptness and setting priorities; important things come first and merit the most time	• Pointlessness of appointments; events begin when 'things are ready'; everything is important
• Projects should be continued without interruption until completion	• No concept of 'closure'
• Cultural reaction time (to threat, challenge, insult, injustice) relatively long; importance of 'lead time' and 'adumbrative sequences' permitting recognition of upcoming events and reactions 'in time'	• Cultural reaction time relatively short; greater spontaneity
• Low ability to recognise subconscious rhythms and subtle changes in their lives and social environment	• High ability to recognise and adjust to rhythms of their lives and subtle changes in their social environment
• Patterns of time use may be deeply symbolic revealing status, importance, responsibility, of social actors and events, e.g. leeway in time use as symbol of authority, time of day (early morning, night) as symbol of urgency, communicative significance of being late and 'waiting time'	• Patterns of time use are not symbolic but dependent on more complex social processes
• Business is conducted confidentially, outsiders are screened out; individualised relationships	• Business is conducted in public among groups
• Agreements and rules hold	• Nothing seems firm, changes may occur right up to last minute

Culture - A Force of Social Cohesion, but also of Exclusion

Linton[53] notes a number of factors pulling individuals in a society or culture towards a similar pattern of behaviour. For instance, besides directly and indirectly transmitted values and attitudes and corresponding systems of reward and retribution, *common experience* unites members of one social or cultural entity more with each other than with members of other units. Fates of members of any group are, to differing extents, inextricably bound together. History offers myriad examples for both national societies and for cultures within or extending across national frontiers. These are the grounds for a sense of affiliation, allegiance, and solidarity linking mutually unacquainted members of a cultural group and distinguishing them from 'outsiders'. Nevertheless, even within cultural groups members of the human species still have greater potentialities for differentiation and individualisation in behaviour than those of any other species. Thus, in spite of the infinite number of variations in individuals, 'cultural patterns' do evolve in accordance with standards or norms of behaviour and opinion. These cultural patterns are regarded by cultural anthropologists to be building blocks of culture as a whole and are the basis for estimating probability that a member of a given society will react in a certain way. In fact, Linton[54] considers that individual personality or personal predispositions are not revealed by culturally patterned responses but rather by his or her deviations from given cultural patterns.

A society's or culture's modal personality will be expected to exhibit a certain degree of behavioural consistency. Analysis of individual behavioural consistency that persists in spite of differences in stimulus makes use of the concept of 'response sets'.[55] A frequently studied response set is the tendency to endorse socially desirable items. Experiments by Asch, Sherif, Festinger, Back,[56] and others revealed the power of pressure in cohesive groups to conform in behaviour to expectations of others and to accept information as accurate.[57]

Stability of socially derived norms about the nature of the socio-economic environment is related to a lack of awareness of the existence of such norms. If an individual becomes aware of a mediation between sensory evidence and social perception, he searches for further evidence. Cultural interaction thus weakens power of social norms on perception because it provides a wider range of reference.[58] The more obvious discrepancy between sensory and social information becomes, the greater the need for resolving cognitive dissonance.[59] Such discrepancies are also very likely to be noticed by 'outsiders' who bring different scales of reference with them. Destabilisation of socio-cultural norms and behavioural patterns and enlargement of ranges of reference and of perceptual scopes are circumstances increasing cognitive dissonance as well as pressure toward change, innovation, and enterprise.

Differentiating Between Cultures and Nations

The most frequent form of cross-cultural comparisons are those between populations of different countries. This is quite natural because, in most cases, data are surveyed and analysed in aggregates on the national level and it is still at the national level where policy relevance is strongest. National entities have also been shown to be major frameworks for creating and maintaining institutional structures and processes, thus being a major constituting factor of society and, through such institutions, of culture. It is the state which either controls directly or issues mandatory guidelines for formal institutions, such as child nurseries, schools, institutions of higher learning, having, among others, the express objective of individual socialisation through inculcation of socially desirable values, norms and role models. Is differentiation of study populations according to national origins or citizenship a valid method in cross-cultural analysis - particularly in the context of a global march towards, as a whole, superficially more similar, but internally 'fragmented societies'?[60]

Benedict,[61] who introduced the concept of 'cultural configurations' underlying modal personalities in society, acknowledged skepticism about the application of anthropological techniques to modern, civilised nations due to the lack of cultural homogeneity within national borders but she claimed that under these circumstances anthropological studies should aim at ascertaining which attitudes and convictions the various classes and subcultures hold in common. It is not our purpose or intention here to perpetuate national stereotypes or to use previous empirical findings on differing incidences of modal personality types in various nations as a basis for hypotheses on relative strengths of national dispositions toward innovation and enterprise but it is nonetheless important to note that Inkeles and Levinson, and Milgram were able to cite a number of empirical investigations looking for and, in part, confirming the existence of significant differences in certain relevant personality traits between national samples.[62] The question remains however, whether these differences will persist in the context of increasingly globalising and internally fragmentising societies.

Inkeles and Levinson[63] define national character as a function of the regularity with which certain personality patterns (values, behaviours) manifest themselves in individuals belonging to this society. Again, Linton's view of national character is based on the 'modal personality' concept, which does not preclude a wide latitude of variety in individual personality characteristics.[64] This modal personality will most likely incorporate the set of characteristics most congenial to the prevailing social norms and institutions of the country in question, thus reinforcing society's coherence. Inkeles and Levinson[65] pointed to observations by social psychologists like Linton, Kardiner and Fromm that, particularly in swiftly changing modern industrial countries, standards for socially desirable characteristics may evolve faster than current modal personality structures (which are *per definitionem* relatively enduring characteristics), creating particular tensions and reactive mechanisms. As stated, these are circumstances conducive to innovation and enterprise.

According to Inkeles and Levinson,[66] for complex industrial and post-industrial society multimodal concepts of national character are most suited to economically diverse, multiethnic and multicultural societies in today's globalising but fragmented societies. Hence, a specific trait constellation would hardly be found in 60 to 70% of a national population; it would appear more likely that a national character defined by five or six traits might have some of these traits represented in between 10 and 30% of the population. Thus, there is nothing unique about national character and there is little gain in attempting to define specific national characters. Of greater relevance for national comparisons (than modal types based on personality traits of social actors) are typologies of political dispositions having evolved in response to institutionally determined sets of opportunities and constraints.

In their international comparative analysis of roles and identities of individual actors within the context of differing political systems, Almond and Verba[67] used the term national 'political culture', instead of 'national character' or 'modal personality', when referring to specifically 'political orientations', i.e. internalised attitudes (allegiance/apathy/alienation) toward the political system and given standards of political behaviour, based on:

- *cognition*; i.e. knowledge or beliefs about the political system, its roles, incumbents of these roles including the self as a political actor, and system inputs and outputs,
- *affective disposition*; i.e. feelings about the political system, its roles, personages, and performance, and

- *personal evaluation* combining the information and feelings with norms and values to form specific judgements and opinions.

The authors looked for national specificities based on the population's general level of knowledge about society and the political system and on their self-perception in their roles as socio-political actors, in particular with respect to their 'subjective competence' or ability to exert political influence. In analysing the contextual political systems, they differentiate between three classes of variables:

- roles and structures;
- role incumbents; and
- policies, political decisions and their enforcement.

Nations are thus political entities with the overall most potent influence on institutions forming modal sets of values, attitudes, and predispositions. It is their political, not their ethnic component which is of interest for our topic. Since, in foregoing chapters, we noted that the given political system is a key element in socio-structural features of the environment with respect to the evolvement of innovation and enterprise, national comparisons of these phenomena should attribute greater significance to political-institutional aspects in explaining differences between study populations.

Cultural Impacts on Organisations, Innovation, and Enterprise

Glade[68] was one of the first to consequently describe the entrepreneur as a decision-maker operating within a specific social and cultural setting, which he termed 'opportunity structure'. Any formal organisation or firm is an institutionalised network of people interacting with each other in what Schein[69] calls a specific *organisational culture*. He enumerates a number of problems which must be resolved by collaborators, obviously in accordance with the organisational culture which emanates from the specific personality mix within the organisation and from its specific context of prevailing social and cultural norms. For instance: The core mission and other manifest and latent functions of the firm must be accepted by collaborators. Concrete goals must be set and strategies (ways and means) for attaining them must be devised with a certain degree of consensus (division of labour, hierarchy, reward system). Information and control systems must be implemented. There must be a common language and use of terminology. Members must accept group boundaries and criteria for inclusion or exclusion.

Stinchcombe[70] indicated that creation of new organisations involves establishment of new roles and relationships, very frequently between strangers. He proceeds to elaborate the point that cultural traditions in which obligations and loyalty to kin and friends invariably override obligations to strangers limit organisational structures to the pool of kin and friends. This is, in Stinchcombe's opinion, definitely not conducive to efficiency in highly specialised economic processes of resource acquisition, resource management, and marketing. He finds that universalistic institutions and values and attitudes are much better suited to constructing social systems out of strangers. But Stinchcombe is obviously referring to already existing, large-scale, highly specialised firms. As we have elaborated in the foregoing, new ventures and SMEs of particular importance for sustainable regional development can take advantage of substantial reductions in 'transaction costs', of increased flexibility, and of enlarged webs of potential financial and informational relationships when they are family-based.

How does the socio-cultural setting affect the individual's disposition to enterprise and innovation? Shapero and Sokol's model[71] describes the push and pull exerted by social and cultural factors on life-path changes, perceptions of desirability and feasibility, etc. leading to new firm formation. We noted that Shapero[72] had himself previously emphasised the social marginality of entrepreneurs as being persons having failed in traditional social and economic roles. He asserts that research data show that individuals are much more likely to take entrepreneurial initiatives in reaction to negative information or events. 'Negative displacements are found to precipitate far more company formations than do positive possibilities.'[73] Brockhaus[74] found in a sample of entrepreneurs that 59% of them had desired to start a business before they had a product or service idea, as compared to 14% having had the idea first. This would seem to support the notion of the entrepreneur being pushed. Similarly, Hagan[75] regarded 'status deprivation' as a stimulus for innovation. Hisrich[76] found that the push away from frustrating experience at work was a particularly strong motivating factor for female entrepreneurs. In addition, if we accept that entrepreneurship involves a certain element of aggressive behaviour, then the fact that every society has its specific structure and patterns for the expression of aggression[77] is revealing with respect to cultural influence potential on entrepreneurship.

Whether the relationship is positive or negative, intentions to innovate, found a firm, etc. obviously depend to some extent on culturally prevalent norms, attitudes, beliefs, and perceptions. However, general normative standards, etc. do prove to be somewhat less influential in persons with internal loci-of-control,[78] such as is attributed to entrepreneurs. Nevertheless - as substantiated by evidence presented in part in a foregoing chapter and to be further substantiated in an upcoming chapter - influence exerted by role models in personal networks of entrepreneurs, such as by parents and 'significant other' friends, is considerable.

Groups alleged to display a particular propensity to entrepreneurship do so only under specific cultural and historical conditions, for instance after immigration to the environment engendering entrepreneurship. Prior to this situation these groups are mostly indistinguishable from those around them.[79] Probably, clashes of ideas between cultural systems, between subcultures and classes within a society, between families or other groups within a community or between individuals are important sources of change. Hence, societies encouraging mingling facilitate a fruitful 'conjunction of differences'.[80] Another factor of importance according to Barnett[81] is the innovator's intellectual perspective, i.e. whether his or her horizon is limited to knowledge existing within his or her society, or whether it extends well beyond to knowledge possessed by adjacent or distant cultural groups. Innovation is impossible for an individual beyond limits of his understanding or his experiences.

Cultural anthropologists like Linton[82] note that, to be accepted in a society, a new thing (an innovation) must be compatible with the modal personality of that society's members; the only way to compare collective dispositions to innovation and development is to study individuals in different societies and cultures. Busch[83] arrived at similar conclusions arising out of his studies of innovation diffusion. Lifeworlds of other cultures should be taken seriously, in particular nonrational aspects in the population's comprehension of innovative ideas and in the adoption processes. Metaphors and similes used in communication by the agents of change, gatekeepers,[84] and opinion leaders in different cultures are of particular interest. Motives and interests of actively and passively involved institutions and power elites are essential factors.

Barnett[85] points out that innovation flourishes in cultural settings anticipating it, especially when expectations in change have positive connotations, such as the notion of 'progress' after the Enlightenment in Europe. Freedom is an important social asset for the innovator who tends to be an individualist. Competition may be a potent incentive for innovation if rewards are allocated on the basis of performance. Monopolistic practices obstruct competition and are often initiated just for that purpose.[86]

Societies where humiliation is not attached to failure but where recognition is accorded to success are particularly fertile grounds for innovation.

In some societies, individualism is at a premium; in others, emphasis is placed on the collectivity as the legitimate base of decision-making and behaviour. There is no consensus in scientific literature with respect to the effect of collective forms of decision-making on propensities for risk-taking. Whereas Whyte and others found that outcomes of conferences and meetings in bureaucratic organisations reflected inhibition of boldness and risk-taking in decision-making, Wallach, Kogan, and Bem found just the opposite: group interaction and achievement of consensus eventuated in willingness to take higher risks, possibly due to decreased individual responsibility.[87]

The extent to which norms for interaction between members of a society have been institutionalised varies and is usually seen to increase as a function of the level of economic development and of population density. Baumol[88] points to numerous institutionalised constraints on entrepreneurship, ranging from the growing body of laws, regulations, and administrative actions imposing limits on certain activities, to bureaucratic procedures. Some of these constraints are historical and have a natural bias towards favouring existing networks. Baumol contends, however, that frequently these constraints are intentional; being imposed by those threatened by pressures of competition.[89] In short, he believes that the very success of innovative entrepreneurship stimulates imposition of such constraints by runners-up.

Whether at all or the degree to which gender differences in entrepreneurial capability are in-born or culturally imparted is still debatable. Wallach and Kogan[90] alluded to considerable evidence in psychology that women were less capable in dealing with problems/decisions on quantitative matters and spatial relations, while they were more adept than men in judging other people. In addition, they produced evidence suggesting that women show greater conservatism under conditions of uncertainty but, when they are more sure of themselves, are more prone to strong/extreme judgements. If valid, these differences would represent differing personality assets for entrepreneurship and innovation. However, the findings, even if correct, may apply only to the society which was under investigation (USA). Societies throughout Europe have been traditionally patriarchal. Female entrepreneurs are noticeably confronted by culturally determined problems in producing collateral, obtaining credit (support systems, sources of funds), and in being taken seriously.[91]

Barnett[92] elaborates reciprocal relationships between innovation and culture. The cultural setting, as the background of innovation, represents an accumulation of ideas forming the point of departure for development of innovative ideas building upon the past. Barnett notes that some societies provide more stimulation for change by offering communication media and opportunities for concentrating, disseminating, and perpetuating information and ideas (including institutions of learning, libraries, publication and conference centres, apprentice systems) and for collaboration while others inhibit fruitful communication and isolate individuals by establishing institutions encouraging secrecy, reglementation, and exclusiveness. Thus, favourable cultural circumstances, similar to favourable socio-structural conditions discussed above, reinforce innovation and enterprise and, in turn, are themselves amplified by the dynamic processes they engender. These are the socio-cultural prerequisites of 'learning regions'/ milieux innovateurs.

Power Structures

Power is basically the ability to exert influence. Wolfe[93] describes power as the potential to induce forces toward desired behavioural movement or directional change at given points-in-time. On the other hand, Stotland[94] describes political or social 'power' as the authority to prevent others (in inferior positions in the power hierarchy) from reaching his or her goal. According to Cartwright,[95] a major base of an agent's influence or power is the possession or control of resources in the sense that their use or allocation can be used to either facilitate or hinder goal attainment by other social actors. Thus, both variants of actions, positive and negative, are manifestations of power.

How do power relations function? Types of power according to Collins and Raven[96] are:
- *informational power* (being or controlling a source of information and communication),[97]
- *coercive or rewarding power,*
- *power of reference* (being a social model or reference group),
- *expert power* (enjoying trust in competency),
- *legitimate power* (entrusted with authority), or
- *negative power* (being a negative model).

On the receiving end,[98] the social actor may respond by
- *compliance* with the will of the superior agent, e.g., due to his or her vested authority, in ignorance of other alternatives, in the hope of some favourable outcome, or in fear of sanctions for non-compliance,

- *identification*, i.e. self-association with an admired social partner and maintenance of satisfactory role-relationships with superior agents, and

- *internalisation*, i.e. acceptance of influence due to its congruency with the actor's own value system or with prevailing norms and values.

The alternative is, of course, conflict.

Interlocking directorates of large corporations and financial institutions, often including prominent members of political parties and incumbents of political offices, and the crucial role of financial and political institutions vis -à-vis existing enterprises and new ventures are evidence enough that in questions of economic development 'power relations' cannot be neglected.[99] It has frequently been pointed out that monostructural and oligopolistic regional economies are inflexible and particularly endangered in times of recession. In addition, oligopolistic power structures have been noted to create severe disadvantages for 'new and innovative entries'. In regions with oligopolistic structures, local business cultures will tend to be dominated by management of the major enterprise(s) with a variety of consequences:[100]

- An entrepreneur's status in the community will be defined by his or her position within the hierarchy of the dominating firm(s).

- The major firm(s) will most likely be paying better and offering more job security than other sectors or new ventures. Recruiting and retaining qualified personnel will then be difficult for 'outsiders'/ 'newcomers' and the regional presence of traditional skills and crafts will decline..

- Capital in the region may be absorbed by the main corporation(s) and transferred to their branch operations elsewhere rather than being put into local banks and financial institutions granting loans to small, innovative enterprises in the area.

- Access to material production resources, to information, and to distribution channels will be dominated by the major concerns.

- Social and economic prominence of the oligopoly's leadership and participation by managers in local and regional politics (e.g. being elected to office, lobbying, donations) will tend to distort the local political culture. The corporation's problems will be kept at the top of the political agenda. Industrial, commercial and other professional associations in the area will be primarily concerned with the needs of dominating entities, depriving the rest of the regional economy with innovative leadership. Innovative ideas for unrelated products and services are unlikely to evolve in such environments. The composition and qualifications of the regional labour force may differ significantly

from the diverse needs of new-born, small, and innovative enterprises. and
- The establishment frequently succeeds in keeping out unwanted 'outsiders'/'newcomers', i.e. those disturbing established regional balances. Fear of inaccessibility to political power in less diversified local and regional economies acts as a deterrent to 'outsiders', thus hampering a diversification of regional or local economic structures.

Disposition over resources or decisional competence with respect to, for example, public resource allocation is a very significant kind of power much coveted by economic elites and is a driving force in local political culture. According to Casson,[101] goods permitting multiple access are 'public goods', an important example being information; one person's access to it does not, in principle, impede another person's access. However, free access to public goods is in many cases only relative. Certain public goods can be conceived as having only limited capacity, beyond which congestion occurs. When access to public goods has limited capacity or a limited number of sources, then it must be bureaucratically administered. This may well be the case with specific types of information and certainly is the case of public funds. Gibb[102] notes that there is, for instance, a wide variety of programmes and projects in the European Union supporting SMEs, but major areas of EU expenditures in science and technology still largely reflect priorities of major European firms.

To quote Linton:[103] 'no matter how inurious an existing institution may be to a society in the face of changing conditions, the stimulus to change or abandon it never comes from the individual upon whom it entails no hardship. New social inventions are made by those who suffer from the current conditions (or by those who stand to profit from an innovation - author), not by those who profit from them.' Hence, regional economic and political oligarchies will, by their very nature, be opposed to any change upsetting the balance which they have established. When protagonists in local political culture consist of corporate managers, politicians, and civil servants, as is usually the case, then we have a local power hierarchy dominated by bureaucratic, organisational man bent on either maintaining the status quo or permitting only gradual, incremental change which can be managed by already existing institutional structures and power hierarchies.

Hence understandably, Gatewood et al.[104] suggest that entrepreneurial activity is most likely to occur under conditions described in conflict models of community power in which vested interests are defended by power elites well integrated into existing associational or institutional networks. According to these authors, a number of studies point to differences in power structures and in sharing and exercising of power between small, rural communities and urban communities. Small, rural communities have stronger self-defence mechanisms against disruptive change. More personal and multifaceted interaction between socio-economic actors in small, rural communities allegedly increases individual participation in decision-making, leading to an atmosphere favouring cooperation among locally accepted players more than competition, especially with outsiders. We recall however, that the dynamic entre-preneurial personality is not particularly compatible with cooperation or partnership models of social behaviour but tends rather to competition and personal achievement. Gatewood et al. thus conclude that while conflict models of power are more likely to encourage entrepreneurial activity, small, rural communities tend to favour stability, stifling initiatives of non-members of power elites.

Promoting regional development through existing structures at the grass-root level would thus have a general tendency to inhibit, rather than to induce, innovation and creation of new enterprises by those outside the establishment. Formulation of subsidisation policies by existing business interests and political administration and channelling supranational and state funds for regional development through well established institutions or agents of a particular sector or interest group have a net effect of rein-forcing existing, sometimes otherwise nonviable structures to the compar-ative disadvantage of 'outsiders'/'newcomers' and innovative noncon-formists. Funding of measures for rural development and research by the European Union within agricultural programmes is a case-in-point.

Social Networks and Peer Groups

Within society there are two intersecting groups of primary importance for the individual's social roles, social identity and socio-economic status: his or her ego-centred 'social network' and 'peer groups'; these are the social actor's major groups of reference. Since these groups are not formally institutionalised but rather are highly personalised, they and the relationships they engender are, by nature, 'informal' and of special significance for nascent enterprise and innovation. Patterns of social relationships with members of these groups as well as prevailing attitudes and expectancies within the groups condition human behaviour and self-perception. Members of *social networks* are active sources of (most frequently 'informal') material and immaterial support.

Barnes[105] described social networks as consisting of 'significant' direct interpersonal contacts and indirect contacts through one or more intermediaries. These contacts are selective, representing only a fraction of all possible contacts within the social environment. Network members may be linked by flows of information or other immaterial or material resources and by similarity in attitudes and motivations affecting their behaviour. Links between the individual and members of his or her ego-centred social network may well be vertical and in imbalance with respect to socio-economic stratification and to the direction of material and immaterial support. Figuring prominently as particularly strong elements of social networks are, in the usual order of importance, members of the individual's family household, extended family, circle of friends and neighbours, colleagues, and casual acquaintances, associates, etc.

Peer groups have a passive but powerful role as models for self-assessment and motivation. Criteria for identifying members of peer groups are more difficult to define. Strictly defined, ego-centred peer groups are a nexus of persons of equal standing or rank or having a status to which the individual aspires. They must be visible to the individual who identifies him- or herself with them in some manner. Members of peer groups are most likely to be considered as matches, mates, or rivals; peer groups are thus primary groups of reference and 'unmapped' sources of motivation. The strong presence of entrepreneurs in a region enhances the visibility of enterprising and innovative role models.

Logically, an intersection of these groups (i.e. of social networks and peer groups) is predisposed since, as a rule, many friends, colleagues, and relatives may well fulfil several criteria. However, many persons considered to be peers may be personally totally unknown to the individual in question and visible to him only through mass media or even hear-say.

Moreno's sociometric analysis of relationships in their geographic and social contexts was one of the initial steps (1934) taken in network analysis. Rippl[106] differentiates between the

- 'emotional support network' in which norms, values and attitudes are often exchanged and reinforced;
- the 'social support group' which assures functional exchange and support with goods and services; and
- the 'informational support group' which functions as a communication network.

Network analysis could, of course, be extended indefinitely to include innumerable elements. Quoting Aldrich and Zimmer,[107] 'a central interest of network theorists, therefore, has been to find ways to set meaningful limits to the scope of a social unit under investigation. The concept of role-set, action-set, and network provides us with some tools for setting such boundaries. A *role-set* consists of all those persons with whom a focal person has direct relations. Usually the links are single-step ties' and are status-related. For entrepreneurs, the authors cite as examples: partners, suppliers, customers, venture capitalists, bankers, other creditors, distributors, trade associations, family members 'An *action-set* is a group of people who have formed a temporary alliance for a limited purpose.' A *network* is defined by the authors as the totality of persons connected by a certain type of relationship.

Aldrich[108] stresses the importance of brokerage roles in innovativeness of networks. Brokers link actors having complementary interests, transferring information, and otherwise facilitating interests of actors not directly connected to one another. 'Brokers' are central persons in networks because they link persons having complementary interests while minimising transaction costs. Thus, Aldrich and Zimmer[109] propose that rates of entrepreneurship should be higher in highly organised populations, i.e., voluntary associations, trade associations, public agencies, etc. should increase probability of fertile connections. On the other hand, although advantages of high density networks for resource consolidation, etc. were illustrated above with respect to competitive advantages of family-supported enterprises, Granovetter[110] produced evidence that people with more diverse role-sets and 'weak' connections via brokers, etc. have access to a wider range of specialised information - a decisive factor in entrepreneurship. It is apparently difficult to establish just how social networks and peer groups function with respect to innovation and entrepreneurship. That even their 'weak' components are an important factor has been well established in empirical research[111] is beyond question. They certainly would be a worthwhile study object for research on innovators and entrepreneurs.

In long-standing networks, communities or groups (especially families), we have noted that transaction costs decline as a result of hard-to-discern relationships of amity, obligation, etc.[112] *Homo oeconomicus* cedes to *homo sociologicus*.[113] Schein,[114] too, indicated that long-standing and well functioning interpersonal communications among social actors tend to maintain stable social relationships, roles, and identities and, by consequence, considers them to be one of the most potent forces *against* change, such as the creation of a new venture or innovation. Stability is highest in the context of multiplex socio-economic relationships within social networks, such as can be found particularly in rural areas.[115]

As mentioned in discussing the relevance of personal traits for innovation and enterprise, using Rotter's 'internal/external locus-of-control scale' Shapero had found his entrepreneur sample to have a very strong internal locus of control. However, in differentiating his sample according to (two) national groups he found differences in the degree of self-reliancy/ autonomy. He attributed this result to different prevailing attitudes in the respective societies in which the groups were embedded. The major factor for all entrepreneurs he studied was differing functionality of informal networks in national samples through which information and judgements on possible 'deals' were circulated outside of conventional structures and institutions. Shapero's conclusion was that communities lacking informal networks are not likely to develop and retain new companies or to attract individuals who might create them.

Birley[116] studied the role of networks in founding new firms, differentiating between the social actor's 'informal networks' (family, friends) and 'formal networks' (banks, accounting firms, law firms, various associations). She also found that informal networks were a powerful influence in new venturing whereas the role of formal networks was negligible. Furthermore, in studying networks for already existing firms Birley once again differentiated between the entrepreneur's 'formal networks' (banks, accountants, lawyers) and 'informal networks' (family, friends, business contacts). Her survey showed that, here too, informal networks played the primary role in financial support of small firms; banks only being approached near the end of set-ups when most resources had already been informally organised. Mingione and Bagnasco[117] point to the example of the 'Third Italy' (Northeast and Central Italy) as an indication that family farming traditions making extensive use of informal relationships provide a wide spectrum of attitudes and 'reciprocal' relationships in family and social networks which are particularly amenable to initiation of new businesses.

Reference groups are important for the individual's formation, maintenance, and change of attitudes and for his or her social identity or status.[118] Homophily is the degree to which pairs of interacting individuals have similar attributes, such as beliefs, values, levels-of-education, social status, etc.[119] The family is a densely linked social network in which, as a rule, there is, due to a comparably high degree of trait homogeneity and intensity of relationships among its members, strong coherence in political attitudes.[120] As repeatedly stated, this reduces enormously transaction costs.

People at any particular social or professional level tend to most fre-quently come into contact with their peers.[121] Kourilsky[122] concludes from her study of relevant literature that individuals who rely heavily on peer reinforcement (in the sense of *moral* support) are more suited for non-entrepreneurial positions. An internal locus of control remains being considered a main defining characteristic of entrepreneurs. However, Stotland's experiments corroborated the hypothesis that supportive peer groups serve to heighten persistence toward one's own goals and aggress-iveness in the face of a threatening power.[123] Even strong-minded entre-preneurs are appreciative of moral support from respected peers.

Shapero and Sokol[124] claim that the larger the number and variety of entrepreneurs in a particular culture, the greater is the probability that individuals in that culture will form companies. Besides members of one's own family household, other kin, friends, colleagues and classmates, etc. offer credible peer models to potential entrepreneurs. Cooper[125] found, for example, that high-tech entrepreneurs tend to locate themselves in the same area as their previous employer (their incubator) and to further develope products closely related to their previous activities. Logistical advantages and lower transaction costs are only part of the story; existing networks are a crucial factor. A strategy for regional development has evolved since the early 1980s in the U.S., Western Europe, Japan and China aiming at creating 'business incubators', i.e. centres functioning as brokers linking potential business partners or as communication forums to facilitate constructing and maintaining networks.[126]

Peer influence in willingness-to-innovate is also strong. Rogers and Shoemaker[127] found that 'early adopters', i.e. those 'quick and easy' receptors of innovation, have a higher intensity of social participation and are more highly integrated into the social system (networks) than 'late adopters'. Roger and Shoemaker's 'early adopters' are more cosmopolite and have greater exposure to interpersonal communication channels. They more actively seek information about innovations and thus have superior knowledge about them. They also exhibit a greater degree of opinion leadership; are less dogmatic and fatalistic. Paradoxically, with respect to their role as sources of information, Rogers and Shoemaker[128] did also recognise that, very often, the most innovative members of a system are perceived as deviants and are accorded a dubious status with low credibility by most other members of the system. Their role in *actively* persuading others to adopt the innovation is likely to be very limited but it is their example which may engender imitation by others. Hence, it is quite apparent that there is a distinct difference between genuine 'innovators' and even 'early' adopters;   the latter being more subjects of innovation processes than active agents of innovation.

Active propagation (or prevention) of innovations can either be a function assumed by 'gatekeepers' and 'opinion leaders', i.e. those exerting informal leadership (but not themselves 'innovators'), or by 'agents of change' (e.g. 'brokers', local developers, politicians) who have the task of promoting the change in question. *Gatekeepers* open access to networks. An *opinion leader* is a key person in social networks or peer groups to whom others most often (but not invariably) turn for information. He or she is able to influence informally other's attitudes and overt behaviour in a desired way with a relatively high frequency. Their function is purely informal and not tied to any position or status within the system. They prove to be only relatively more influential than others and are themselves highly subject to influence. At the turn of the (twentieth) century, Tarde[129] contributed valuable insights into the process by which behaviour of opinion leaders is imitated by others. Gould and Törnquist[130] cited several studies demonstrating the important role of opinion leaders in triggering the adoption of innovations.

Robertson[131] provides a profile of opinion leaders as contrasting to their followers. They resemble very much their followers, generally having the same social status, but would appear to be somewhat more cosmopolitan. They are highly gregarious, are more exposed (at least in their field of 'expertise') to mass media than their followers and exhibit a greater knowledge about the field in which they are considered to be an opinion leader. They are much more receptive to innovation than those who value their advice but are themselves generally not innovators. They show a greater adherence to prevailing norms than do their followers. Thus, they would appear to conform to ideal social models. Since they conform relatively strongly to prevalent social norms and enjoy a relatively prestigious status within the given social structure, it is evident that opinion leaders have a basic interest in preserving the status quo unless they recognise benefits in altering regional balances. Opinion leaders can thus be instrumental in either promoting or preventing the spread of innovations. This characterisation of opinion leaders closely corresponds to that by Rogers and Shoemaker[132] who describe relationships between opinion leaders and their followers as a function of homo- or heterophily of the social partners. If partners are dissimilar, followers would tend to seek opinion leaders of a higher social status, with more education, with greater mass media exposure, with more cosmopolitan ways of life, with more experience in innovation, and with more expert contacts.

Figure 1 proposes a simple classification scheme for identities of social partners of innovators and entrepreneurs, for instance for analysing time inputs into different types of communication and joint activities with persons belonging to their ego-centred networks or peer groups.

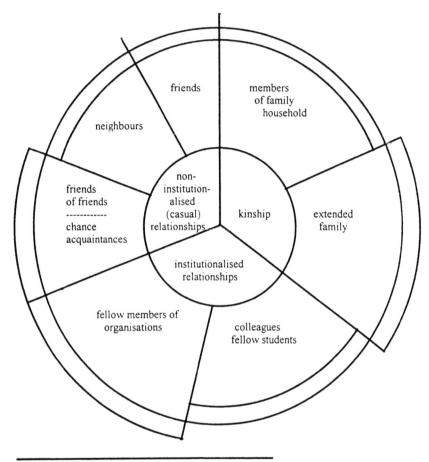

eye-to-eye communication on a regular (daily/weekly) basis (inner ring)

communication mainly by mail, electronic media (outer ring)

Location of social partner's residence: - immediate vicinity of respondent's residence
                                        - in respondent's neighbourhood
                                        - in respondent's community/area of town
                                        - in a nearby community (within a half-hour reach)
                                        - further away/very remote

Communication and Information Flow

According to Lasswell and McGuire,[133] communication processes can be divided into five components: source, message, channel, receiver, and destination. Similarly, as noted in the aforegoing, Rogers and Shoe-maker[134] had defined *communication* as the process by which messages are transferred from a source to a receiver, differentiating in their model *source, message, channels* of transmission, and *receptors. Diffusion* they defined as the process by which innovations (messages) are spread to members (receptors) of a social system. Little of substance is known about the interrelationship between communication structures and media, on one side, and entrepreneurship. Two areas of communication research would appear at least to be on the way to making a contribution to this topic: Research pertaining to effects of telecommunications on economic globalisation, on one hand, and on creation of new ventures in rural areas, on the other. However, the impact of communication structures and media on innovation diffusion has long been a topic of economic geography, communications research, and rural sociology.

According to Katz,[135] until about 1960 communication studies investigating the proliferation of new ideas, etc. assumed that society was com-prised of a collectivity of atomised individuals subject to influences of mass media. In the following decades, research has indicated, however, that mass media's influence on their targets was far less potent than had been expected. Even in the age of mass communication, social geographers like Hägerstrand[136] also stressed importance of neighbourhood effects, i.e. of informal links between friends and acquaintances for directing information and influence. This then came to be the golden age of rural sociology since several representatives of this discipline had also been investigating the proliferation of new ideas and farm technology in agricultural areas. Having conducted many of their empirical studies in small, isolated communities, they did not assume that their respondents did not speak to each other, as their colleagues in communications research had implicitly done. In fact, the existence of interpersonal relationships in groups of family, friends, neighbours, and co-workers significantly intervening in information flows was a major premise of rural sociology. Was not Tönnies' Gemeinschaft a defining characteristic of the rural as opposed to anonymous Gesellschaft in urban settings? When commun-ication research had learned this lesson from rural sociology, the door was opened to the concept of 'two-step flow' of information.[137] Rogers[138] considers that impersonal sources of information (e.g. mass media) are most important for becoming aware of innovations - the first of the two steps, whereas person-to-person contacts play a stronger part in evaluating their actual benefit to the innovator - the second step. Katz[139] stresses persisting strong significance of person-to-person contacts in communication, even in the age of modern mass media. He also sees mass media as being most important for the initial 'awareness' stage of information processing, while considering the innovation acceptance stage to be more an outcome of interpersonal communication processes. Hence, personal influence appears to be ultimately more effective than mass media in gaining acceptance for change. In other words: When decision making is differentiated according to phases, such as becoming aware of an innovation, becoming interested in it, evaluating it, deciding to try it, etc., mass media appear relatively more powerful in early informational phases whereas personal influences are more potent in later deliberation and decision phases. Another conclusion: 'Early adopters' are more likely than others to have been influenced by impersonal sources (institutions and media) outside of their community and at greater distances. They prove to be relatively more open to rational evaluation of changes and to contact with the world outside their own community. By their example, they exert a passive influence within their own communities as part of the second step of information flow.[140]

Lerner[141] emphasises the major role of information flows for distribution of power, wealth and status in society and he proposed a model of communication systems back in the days of relatively unsophisticated mass communications (the 1950's) which still has retained much of its validity; see the following scheme in Table 5.2 based on his model. For Lerner, the direction of change in communication systems in that era was clearly from oral systems to the media.

Table 5.2   **Traits of communication systems according to Lerner**

Trait:	System: Oral	Media
channel	personal (face to face)	broadcast (mediated)
audience	primary groups	heterogeneous masses
content	prescriptive ('rules')	descriptive (news)
source	hierarchical (status)	professional (skill)

Many studies of diffusion of new ideas within and between societies have been undertaken since the 1950s. Some studies have investigated the role of 'gatekeepers' in channelling information.[142] Dosi[143] recognises communication and information flows resulting from cultural and geographic mobility of social actors as being important for the search, development and adoption of new processes and products. Brown[144] states that adoption of innovations is primarily an outcome of learning or communication processes. In his studies of innovation diffusion, he contends, however, that spatial diffusion depends on the type of innovation in question, determining whether agents of diffusion have a mono- or polynuclear (decentralised) propagation structure and whether diffusion processes rely heavily on existing infrastructure networks.[145] Rogers and Shoemaker also maintain that a crucial element in diffusion are characteristics of the innovation itself. Characteristics of innovations not only determine target groups, i.e. groups susceptible to becoming adopters, according to the authors they affect:[146]

- their *relative advantage*, i.e. perceived degree of superiority to superseded items/processes
- their *compatibility* with existing values, past experiences and needs of receptors
- their *complexity* with respect to their understandability and use (affecting channels)
- their *'trialability'*, i.e. their ability to be tested on a limited basis by receptors and
- their *observability*, i.e. the visibility of their outcomes to non-users (another channel).

Festinger[147] points out that the drive to communicate is significantly increased by pressure within groups towards uniformity. Taking into account previously discussed typologies of the entrepreneur revealing inclinations to nonconformism or dissidence, it is probable that entrepreneurs are generally less communicative than others in order to avoid unnecessarily generating conflict or divulging key elements of his or her 'entrepreneurial idea'. Nevertheless, a promising topic of research would be communication patterns of entrepreneurs and of initiators of new ventures. Are there also two-step-communication processes in enterprise, as is attributed to innovation diffusion? Just how important are eye-to-eye encounters in the age of electronic media? The development of global cities would seem to indicate continued importance of direct contacts/access to partners, peers, etc. Törnquist[148] explains this primacy of direct personal contacts by pointing to their flexibility (those involved in information exchange are potentially both transmitters and receivers), to the involvement of several senses giving communicators many instruments of expression, and to common situational contexts of partners in communication. Trust between individuals is still best assured and reinforced through repeated and spontaneous personal encounters. Due to the easy accessibility of a wide range of sources of highly specialised information in agglomerations, Stearns et al.[149] estimate that firm survival chances be higher in urban than in rural areas.

Hägerstrand provided a key concept for the investigation of communication processes in space, i.e. the *mean information field* of periodic contacts, based on the relationship between geographical distance and frequency of contact.[150] Unfortunately, we know of only one data-supported description of specificities of the mean information field of entrepreneurs. According to Sweeney[151] information-seeking behaviour of entrepreneurs is very personal and 'informal', i.e. they rarely go beyond the circle of friends, other personally known entrepreneurs, business associates, or persons recommended by them. We know nothing of their temporal patterns of communications. Studies have been made of professional contacts in other socio-professional groups, for instance: corporate managers, as we have seen - a group sharing some important functions with entrepreneurs. Thorngren[152] investigated 'contact systems' by investigating contacts of 3 000 executives in a three-day-period with a total of 15 000 contacts. Studies of this type would be instrumental in learning more about the grass-roots of entrepreneurship.

Effects of most recent developments in information and communication technology on entrepreneurship have already been addressed in Chapter Two.

Notes

1   Cartwright 1959, p. 3.
2   Aldrich and Zimmer in Sexton and Smilor 1986, p. 4ff.
3   Cf also Malecki 1994 and 1998; and Robertson and Langlois 1995. Halstead and Deller (1997), acknowledging that the presence of good infrastructure is a pre-requisite, but not a guarantee for development of small manufacturing firms, found that 'Quality-of-Life' factors were more important for entrepreneurs' site preferences.
4   Nittykangas 1996, p. 58.
5   Cf. Reynolds in Sexton and Kasarda 1992, p. 269ff.
6   Aldrich and Zimmer in Sexton and Smilor 1986.
7   Gartner 1985, p. 700f.
8   Cf. Shapero and Sokol 1982.
9   Bruno and Tyebee in Kent, Sexton, and Vesper 1982.
10 Kilkenny 1998
11 Cf. Camagni 1991, p. 1-9.
12 Cf. Alves and Morrill 1975, p. 298.

[13] Rogers and Shoemaker 1971, p. 29ff.

[14] Cf. Collins and Raven in Lindzey and Aronson 1969, p. 102f.

[15] Brown, Malecki, and Spector 1976. See also Brown and Cox 1971.

[16] Aldrich and Zimmer in Sexton and Smilor 1986, p. 6.

[17] Cf. Cecora 1994.

[18] Katz, Levin, and Hamilton 1963, p. 248.

[19] Stinchcombe in March 1965, p. 145f.

[20] Cf. MacMillan 1983.

[21] Cf. Pennings in Kimberly, Miles et al. 1981, p. 138 and 152.

[22] Cf. Pennings in Kimberly, Miles et al. 1981, p. 139ff.

[23] Roberts and Jung 1995, p. 53f and 75ff.

[24] Almond and Verba 1963.

[25] The terms used for categorisation: participant, subject, and parochial political systems.

[26] For instance, Italian society was described as being characterised by political and social alienation, i.e. that most Italians seem to decline political and social engagement and have low self-ratings of their 'subjective competence' or ability to change things (p. 402f.).

[27] Shapero and Sokol in Kent, Sexton, and Vesper 1982, p. 73.

[28] Noteworthy exceptions are certain contributions to 'institutional economics; cf. Cécora 1993. We might also allude to the questionable attempt by Bowles (1998).

[29] Whiting 1968.

[30] Cf. Linton 1981, p. 2 and 56; See also Kroeber and Parsons 1958, p. 582.

[31] Cf. De Vos and Hippler in Lindzey and Aronson 1969, p. 333ff.

[32] Our term.

[33] Hall 1982.

[34] Kroeber and Kluckhorn (1952, p. 181) cited by De Vos and Hippler in Lindzey and Aronson 1969, p. 323f.

[35] Cf. Cécora 1994.

[36] Jacobs and Campbell 1961; also cited by Tajfel in Lindzey and Aronson 1969, p. 356.

[37] In her first publication of Patterns of Culture. Boston 1934. Cf. also Inkeles and Levinson in Lindzey and Aronson 1969, p. 419; and Singer in Kaplan 1961, p. 25ff.

[38] Cf. Lindesmith and Strauss 1950.

[39] Cf. Wallace 1952. According to Wallace, individuals react differently to environments, at best, one can speak of probabilities of this or that reaction.

[40] Parsons and White in Lipset and Lowenthal 1961, p. 99f.

[41] Tajfel in Lindzey and Aronson 1969, p. 317ff.

[42] Tajfel in Lindzey and Aronson 1969, p. 359ff.

[43] Hall 1982, and Tajfel in Lindzey and Aronson 1969.

[44] Cf. also Hymes in Kaplan 1961, p. 313ff.

[45] For instance, symbols of politeness and etiquette (e.g. closest permissible 'personal distance', furthest possible 'social distance'), of punishment, encouragement, dominance, aggression, interest, distaste (e.g. facial expressions, body posture, gestures, tone and pitch of voice).

[46] Hall 1973, p. 38ff; and 1976.

[47] Cf. Cochran 1965, p. 27 and 35f.

[48] Cf. Hall 1983, 1973, 1976.

49   Hall, 1983, p. 153-163 and 1973, p. 46.
50   Such as the Truk people in the Pacific.
51   For example, the Sioux Indians.
52   Hall 1976, p. 20.
53   Cf. Linton 1981, p. 14ff.
54   Linton 1981, p. 26.
55   Mischel 1968, p. 83ff.
56   Asch 1951; Sherif 1947; Sherif and Harvey 1952; Festinger 1954; Back 1951.
57   Threshholds of conformity may well vary between cultures, as Tajfel illustrates with Flament's characterisation of French *esprit de contradiction* as opposed to American *esprit conformiste*.
58   Tajfel in Lindzey and Aronson 1969, p. 355 and 345ff.
59   Festinger 1957.
60   Cf. Mingione 1991.
61   Benedict 1946, p. 276f.
62   Inkeles and Levinson in Lindzey and Aronson 1969, especially p 445ff. They quoted e.g. studies by Almond and Verba (1963), Buchanan and Cantril (1953), Cantril (1965), Morris (1956), Allport and Gillespie (1955), and McClelland (1961). See also Milgram 1961. More recent results can be found in the publications generated by the European Values Study.
63   Inkeles and Levinson in Lindzey and Aronson 1969, p. 424.
64   Linton 1981. Cf. also Inkeles and Levinson in Lindzey and Aronson 1969, p. 424.
65   Inkeles and Levinson in Lindzey and Aronson 1969, p. 425.
66   Inkeles and Levinson in Lindzey and Aronson 1969, p. 426f.
67   Almond and Verba 1963 and 1965, p. 11ff.
68   Glade 1969. Cf. also Low and MacMillan 1988, p. 149f.
69   Schein 1983.
70   Stinchcombe in March 1965, p. 148ff.
71   Shapero and Sokol 1982; see also Low and MacMillan 1988, p. 150.
72   Shapero 1975.
73   Shapero and Sokol in Kent, Sexton, and Vesper 1982, p. 79.
74   Brockhaus in Kent, Sexton, and Vesper 1982, p. 51.
75   Cf. Hagan 1962.
76   Hisrich in Sexton and Smilor 1986.
77   Cf. De Vos and Hippler in Lindzey and Aronson 1969, p. 327ff.
78   Krueger and Carlsrud 1993, p. 325.
79   Aldrich and Zimmer in Sexton and Smilor 1986, p. 7.
80   Barnett 1953, p. 46.
81   Barnett 1953, p. 40ff.
82   Linton 1981, p. xvf. According to Linton (p. 4), for studies the individual has been assigned to psychology, society to sociology, and culture to cultural anthropology.
83   Busch 1978. See also Ruttan 1996, p.60.
84   Gatekeepers are (usually senior) members of firms controlling information flow to and from the firm; cf. Macdonald and Williams 1994, p. 123. They differ from 'external (contact) persons', described by Falemo (1989) and others, who channel information and resources to the firm.

85 Barnett 1953, p. 56 and 60ff.
86 Barnett 1953, p. 78f.
87 Wallach, Kogan, and Bem in Cartwright and Zander 1968.
88 Baumol 1982, p. 34ff.
89 I.e. by the 'good old boys' - a term brought back to mind by Frank L. Hefner at the Southern Regional Science Association's meeting in Savannah GA in 1998.
90 Wallach and Kogan 1959.
91 Hisrich in Sexton and Smilor 1986.
92 Barnett 1953, p. 40ff.
93 Wolfe in Cartwright 1959, p. 99ff.
94 Stotland in Cartwright 1959, p. 53f.
95 Cartwright in March 1965, p. 11.
96 Collins and Raven in Lindzey and Aronson 1969, p. 166ff.
97 E.g. power of 'gatekeepers' in information dispersion, cf. Falemo 1989 and Macdonald and Williams 1994.
98 Cf. e.g. Kelman 1961.
99 Cf. e.g. Granovetter 1985, p. 502.
100 Cf. Markusen 1985, p. 280ff; and Sweeney 1987, p. 21ff. See also Freshwater 1996, p. 783; and Simon 1996, p. 8f.
101 Casson 1991, p. 95ff.
102 Gibb 1993, p. 13.
103 Linton 1981, p. 23.
104 Gatewood, Hoy, and Spindler in Hornaday, Shils, Timmons, and Vesper 1984.
105 Barnes in Holland, P.W. and Leinhardt 1979, pp. 403-423.
106 Rippl 1995, p. 80.
107 Aldrich and Zimmer in Sexton and Smilor 1986, p. 11f.
108 Aldrich in Marsden and Lin 1982, p. 288ff.
109 Aldrich and Zimmer in Sexton and Smilor 1986, p. 17f.
110 Granovetter in Marsden and Lin 1982, p. 105ff; Cf. also Aldrich and Zimmer in Sexton and Smilor 1986, p.19.
111 Cf. Sweeney 1987 and Monsted 1993, p. 218. Monsted emphasises the importance of not only studying the (static) composition of such networks but also their dynamics (new entrants, exits).
112 Cf. Aldrich in Marsden and Lin 1982, p. 288.
113 Cf. Cecora 1994.
114 Schein 1960.
115 Cf. Cecora 1994.
116 Birley 1985. Cf. also Low and MacMillan 1988, p. 150.
117 Mingione 1991, p. 57 and 111, and Bagnasco 1977.
118 Cf. e.g. Cartwright and Zander 1968, p. 45ff.
119 Rogers and Shoemaker 1971, p. 210.
120 Lazarsfeld, Berelson, and Gaudet 1960, p. 180ff.
121 Gould and Törnquist 1971, p. 150.
122 Kourilsky 1980, p. 176.
123 Stotland in Cartwright 1959, p. 66.
124 Shapero and Sokol in Kent, Sexton, and Vesper 1982, p. 84.
125 Cooper 1986. See also Low and MacMillan 1988, p. 150.

[126] Smilor and Gill 1986.
[127] Rogers and Shoemaker 1971, p. 188.
[128] Rogers and Shoemaker 1971, p. 34f.
[129] Tarde 1895.
[130] Gould and Törnquist 1971, p. 165f.
[131] Robertson 1971, p. 179.
[132] Rogers and Shoemaker 1971, p. 213.
[133] McGuire in Lindzey and Aronson 1969, p. 172.
[134] Rogers and Shoemaker 1971, p. 11f.
[135] Katz 1960.
[136] Hägerstrand 1965, p. 28.
[137] Katz 1960, p. 439f. Cf. also Cecora 1994.
[138] Rogers 1962. See also Rogers and Shoemaker 1971, p. 13.
[139] Katz 1960, p. 440; see also Katz and Lazarsfeld 1955; Katz 1957; Katz, Levin, and Hamilton 1963, p. 245f.
[140] Rogers and Shoemaker 1971, p. 210.
[141] Lerner 1958.
[142] E.g. Malecki 1998; and Macdonald and Williams 1994.
[143] Dosi 1988, p. 1121.
[144] Brown 1981, p. 6 and 52f.
[145] Brown 1975, p. 185ff.
[146] Rogers and Shoemaker 1971, p. 22f. Cf. also Weiss in Lindzey and Aronson 1969, p. 148ff.
[147] Festinger 1950.
[148] Törnquist 1970, p. 28.
[149] Stearns et al. 1995.
[150] Hägerstrand 1965, p. 30. See also Marble and Nystuen 1963, p. 100ff.
[151] Sweeney 1987, pp. 145, 147, 155ff, and 174.
[152] Thorngren 1970, p. 418ff. Empirical results revealed that the so-called 'programme network', characterised by short contacts of a routine nature having to do with administrating the use of already allocated resources, accounted for 70% of all contacts. The breath of exchanged information was insignificant and the share of contacts devoted to research and development of the enterprise was relatively low. The information flow is mostly monothematic, one-way and well structured. The contact was usually conceived on the day of occurrence and was mainly carried out with the help of technical means of communication. The so-called 'planning network', involving contacts aiming at changing or initiating resource use, exchanges information of a greater scope. The average duration of these contacts is approximately twice as long. Information flow is two-way and occurs both through technical media and face-to-face contacts. The share of contacts devoted to research and development is also low. The so-called 'orientation network', involving interactions of the enterprise with heretofore unknown aspects of its environment and consequently characterised by basically unstructured information flows, exhibits longer-lasting contacts which are more sophisticated in content. It deals with more difficult problems and comprises only 5% of total firm contacts but 80% of new contacts. There are only face-to-face contacts in this network. The planning horizon for such contacts can

easily exceed one week. Information flow is two-way; meetings may last several hours and frequently involves more than three participants.

6 Summary and Conclusions

When mainstream economists leave their 'laboratory' to test economic models on social reality, they are often puzzled to observe effects of intervening socio-cultural factors on 'economic' processes capable of changing the processes themselves as well as their overall socio-economic outcomes. These factors are usually treated as environmental friction on market equilibrium and are integrated as 'error' coefficients in neoclassical models. As a rule, from the macroeconomic perspective, political institutions and circumstances are recognised as (sometimes predictable and controllable) exogenous influences on national consumer, labour, and capital markets. Accordingly, they are instrumentalised for the regulation of resource flow and distribution. However, as attention shifts toward regional economies and microeconomics of firms and households, intervening socio-cultural and political factors increase and compound to the point that it becomes difficult or impossible either to validate economic principles by observing behaviours of study populations or to use them to predict collective outcomes. Planning, implementing, and funding measures for sustainable regional development thus entail taking leave of neoclassical models based on laboratory concepts of populations purported to consist of atomised, individually identical *homines oeconomici* and plunging head first into the social reality of *homo sociologicus*. If specific regional economic mechanisms and circumstances do not fully concord with corresponding development concepts, then congruence of policy goals with regional potentials, and therefore efficiency of development programmes, are jeopardised.

Tailoring development concepts to regional economies is becoming more and more difficult due to globalisation of financial and commercial markets. Globalisation is leading to rapid, largely uncontrolled amassing of wealth and power by international corporate and financial oligarchies which, abetted by neoliberal economic policies, have succeeded in superceding forces of supply and demand on 'free markets' by organising and controlling the markets themselves. Populations living in 'economies in transition' (from socialism) must find it strange that national centrally planned economies are gradually being replaced by a global, increasingly centrally planned economy where the incumbents of power do not even need to organise mock elections to legitimate their control. Other direct impacts of globalisation are:

- concentration of economic and political power in 'global cities', increasing interregional disparities. Strategic global players will locate at strategic sites with optimal transportation and communication networks, remaining in continuous, if possible spontaneous, eye-to-eye contact with each other
- a new valorisation of economic activities creating great disparities between earning potentials of different socio-professional segments of the population. Production of high-value goods and services has become the most dynamic and greatest income-generating sector of the economy. Human capital is the primary resource input into innovative and high-value production and provision of goods and services
- obvious incompatibility between mechanisms of income generation and state instruments for resource distribution. The state has proved to be inadequate in dealing with transnational corporate entities and global financial interests. National states do not collaborate in an effective manner and are individually competing for comparative advantages as global sites. And
- increasing devalorisation and redundancy of dependent, unskilled labour in a global society in which, in the not distant future, one-fifth of the potentially gainfully active population will allegedly be able to produce all the goods and services needed.

How will disadvantaged and rural areas fare in the process of economic globalisation? Will interregional disparities continue to increase or will interregional spill-over effects ultimately counteract global concentration processes? The urgent need to tame globalisation in the interest of sustainable regional employment and development is apparent. It is all too obvious that regional interests are not on the global agenda.

Endogenous and innovative entrepreneurship proves to be a region's prime 'human capital' for sustaining regional employment and development. For some time now, strategies for national and regional economic development have focussed on harnessing endogenous regional potential for innovation and entrepreneurship. How can we identify individuals with innovative and entrepreneurial dispositions and ability? Some claim that modern industrial and post-industrial societies have evolved a dependency culture, i.e., a general preference for dependent employment or, alternatively for reliance on the state for material needs. Others claim that the supply of 'natural entrepreneurship' be constant and equal in all populations and that observable disparities between regions be due to differing regional sets of opportunities and constraints, favourable constellations serving as firm 'incubators', 'learning regions', or 'innovative environments'. Empirical studies have indicated that the pool of potential innovators and entrepreneurs in all populations is greater than is generally assumed. We proposed to show that socio-cultural factors are indeed capable of forming both 'modal personalities' with specific dispositions towards innovation and entrepreneurship as well as environments with specific settings of relevant opportunities and constraints, thus determining relative frequency and strength of manifestations of innovative behaviour and enterprise.

The owner-manager of SMEs, as the 'classical innovative entre-preneur', is the key figure linking global networks in a sustainable way to the region. But owning and managing a firm does not imply the same thing it did decades ago. The innovative, independent 'classical entrepreneur' is no longer the 'classical capitalist'. Indigenous entrepreneurs are those most firmly rooted in the region and are those most willing to bear the burden of material and human 'overheads' (e.g., investments in training of payroll personnel/human capital, process innovations, machine overhaul/ replace-ment, building renovation), in order to assure their flexible and efficient deployment in increasingly specialised production of customised goods and services for world markets, and thus to link regional human capital and other localised production factors to the global political and socio-economic environment. At the same time, they are often very dependent for securing/maintaining credit and market shares on transnational power oligarchies in business and finance. Thus, in elucidating the long tradition of entrepreneur typologies, we have stressed the need, in the context of globalisation, to differentiate more exactly between traits and functions of classical SME-owning and -managing 'entrepreneurs', on one side, and those of other groups sharing some or even many entrepreneurial characteristics or functions, such as self-employed professionals and especially managers of transnational corporations. Corporate managers are frequently mistaken in regional development policies for innovative entrepreneurs. They are revealed as exhibiting many traits of 'bureaucratic, organisational man'. To this purpose, we have resorted to allegorical depiction of two ideal personality types: Our 'key player', defined as 'enterprising, innovative, and independent man', is juxtaposed to 'bureaucratic, organisational man'. Noneconomic factors contributing to generation and creative functioning of the innovative and enterprising personality were reviewed with allusions to this typology.

Empirical observations were shown to indicate that SMEs are important sources of technological innovation and that they do have a certain potential for job creation, even if much of what is considered to be growth of the SME-sector is due to labour market pushes to either liberal professions or to 'independent' subcontracting, both with rather limited job creation potential. Although SMEs can (and should be able to) collaborate in and profit from global corporate networks without losing their autonomy, SME-entrepreneurs were shown to be per definition 'non-global players' with strong material and immaterial ties to the region and hence key agents in sustainable regional development. It became clear that owner/managers of small or intermediate-sized enterprises not only have to deal on a long-term basis with expectations of clients and employees, but also have to cope with volatile, but portentous strategies of global financial and corporate interest groups always in quest of maximum short to middle-term returns on investments and strategic alliances. We also elaborated on the fact that new-born and very innovative enterprises often have to deal with hostile institutional environments. Our analysis of the subject matter formulates a number of premises.

Overregulation strangles innovativeness and, in particular, nascent enterprise in highly developed economies. International corporate management is at a comparative advantage in dealing with it. 'Informal' structures and activities are an important 'testing ground' for mobilising resources for nascent or new-born enterprises. Accordingly, the state should intervene in economic processes (as much as necessary but) as little as possible. State interventions should thus be reduced to the absolute necessary minimum of regulatory activity and to provision of generally accessible facilities and infrastructure, as well as *fora* for information exchange and transfer augmenting and linking human capital in the region. Resource (re-) distribution by the state is *per se* not a productive activity. Although social equity does require to a certain extent redistribution of resources, the basis of taxation must be readjusted to include all significant sources of income. Work income is only one source of income. Financial transactions generating exorbitant windfall-profits should also be increasingly subject to taxation. They should be given a higher priority as a target for taxation than 'informal'/undeclared activity, which often proves to be an important initial 'trial' phase of new ventures not only due to tax and wage 'savings', but also to lower transaction costs and reduced liability for countless regulations.

Resource redistribution in form of subsidies for gain-oriented activity has become a major instrument of regional development policy and local commercial and industrial recruitment. Creation or salvation of jobs in regions with high costs of employee wages and fringe benefits is thought to require some sort of compensation to internationally mobile capital interests. However, reallocation of tax revenue and tax abatements as an incentive for innovative and enterprising behaviour are self-contradictory and counterproductive in view of what we have learned about identity and motives of entrepreneurs. Recent studies even indicate that their effects on business locational decision-making has been largely overestimated. Making profits is, or should be, incentive enough to innovate, to found a new venture, or to establish a new industrial or commercial site. Counterproductive effects of subsidisation policies are twofold. First, under certain but not infrequent circumstances, spending time learning about one's 'entitlements' and how to take advantage of them may already be as income-relevant as allocating time to developing new products, services, production processes or ventures. Subsidies tend to reinforce these unproductive behavioural patterns. Secondly, subsidies tend to grant unwarranted competitive advantages to the 'establishment' and large-scale enterprises over innovative new ventures and SMEs. By favouring already existing, strongly connected enterprises, incentive-dispensing policies reinforce given structures, tend to exclude non-connected 'outsiders', and are frequently counter-productive to creativity, innovativeness and nascent enterprise. They also often incite counterproductive speculation.

Sustainable regional development must be based on productivity of economic activity. Human capital is a universal regional resource with potential for increasing marginal utility. Globalisation does therefore not necessarily relegate rural areas to total insignificance. However, in view of agglomeration effects it must be recognised that prime sites for business, finance, and gainful activity will be located in urban centres attached to global networks. Hence, strategies for sustainable regional development must set their sights on developing or maintaining global urban nodes within reach of the rural population and on establishing durable linkages of key actors to the region. Again, SME-entrepreneurs were shown to be those most likely to remain *in situ*.

Sustainable regional development necessitates social actors' constant participation in economic dynamics implying continuous innovative change and enterprise. Not only must transportation and communication infrastructure conform to modern standards, but training and work processes in the region as well. 'Technological leapfrogging' in regional economies ends, at best, at the currently vulgarised state of technology (t). To be truly innovative and at the forefront of technological development of high-value enterprises, i.e., to advance from t to $t+1$, human capital must already be working at the t-level - a decisive characteristic of 'incubator'-environments. Technological apprenticeship at the current state-of-the-art engenders 'spin-off'-firms, which tend to remain in the region of their 'birth'. Besides formal education, training and practical experience of the indigenous population, regionally prevalent attitudes and dispositions towards innovation and change are decisive socio-cultural and political factors for frequency and success of innovations and births of new firms.

This contribution aimed at highlighting certain inconsistencies and counterproductive features of currently widespread regional development policies purporting to promote entrepreneurship and innovation, especially in rural areas. Flaws in current regional development policies were shown to be, in good part, due to their conceptual dependence on neoclassical economic theory, disregarding 'non-economic' factors. Such factors were examined which contribute to generation and creative functioning of the innovative and enterprising personality, on one hand, and which constitute 'innovative environments' in the sense that they encourage, activate and reward the innovative and enterprising individual, on the other. Our essay makes reference to well established principles of individual and collective human behaviour based on concepts and empirical observations in various areas of social science.

Certain *functions* were shown to characterise the entrepreneur across all specific fields of activity. They are part of his or her role expectations and include:

- recognising innovative products, services, and production services (market opportunities),
- setting goals and strategies for the gainful activity,
- providing financial resources, direct risk- and profit-taking,
- setting up organisational structures and technical means of production ('creative destruction'),
- assuming ultimate responsibility for management and control, delegation of authority,
- competing for market shares, and
- representing the firm, initiating public relations and advertisement.

Although sustaining significant affiliations with their regions, entrepreneurs and innovators nevertheless tend to be outsiders or deviants from the 'establishment' striving to maintain their independence and differ in important traits and dispositions, not only from corporate managers, politicians, and civil administrators, as representatives of the prototype 'bureaucratic, organisational man', but also from many other self-employed individuals. Strong ties to their regional home-base notwithstanding, entrepreneurs, innovators and even 'early adopters' tend to be perceptually cosmopolitan (geographically and culturally) and to be strategically located in wide-reaching social networks and communication channels with privileged access to information brokers and opinion leaders. They make extensive use of a wide range of interpersonal channels of communication and are highly exposed to mass communication media. Our review of entrepreneurship research revealed indispensable *traits* of the 'entrepreneur' to be:

- individualism, including an internal *locus* of control (often displaced persons, misfits)
- disposition to leadership and personnel management
- flexibility and ability to delegate responsibility
- analytical competence, including keen social awareness and social skills, e.g., in communication
- creativity, vivid imagination, endowed with foresight
- receptivity for new ideas, cosmopolitan identity
- experience and self-confidence in the field, practical know-how
- willingness to take initiatives and risks
- ambition, high achievement motivation, need of recognition
- inclination to sober rationality   and
- positive attitude toward and self-identification with change.

Studies in social psychology and cultural anthropology have shown that basic traits do not manifest themselves in behaviour independent of socio-cultural, economic, and political circumstances. They may remain latent until aroused by environmental circumstances.   The individual's embeddedness in cultural norms and institutions, socio-political systems and power structures, informal social networks and peer groups, and especially informal communication structures and information flow all interact with personal abilities to recognise and seize entrepreneurial opportunities or to deal with constraints. *Summa summarum*, empirical findings indicate that personal dispositions towards innovation and enterprise are important but that they must be triggered or liberated by favourable contextual circumstances. The importance of social and cultural embeddedness should not be underestimated.

Influences of culture are not readily recognisable but they affect our perception, motivation, and behavioural patterns. 'Culture' is difficult to circumscribe.   We have made use of Benedict's term 'cultural config-urations' to describe cultural coherence with respect to modal patterns of perception and communication, to formation and transmission of values, attitudes, and dispositions, to structures and functioning of institutions and social groups (e.g., families, social networks, peer groups) and to acqui-sition and allocation of resources like time. Culture is based on common experience and determines feelings of cohesion and exclusion (who is an 'outsider'). Thus, culturally determined values, attitudes, role expectations, and personal identities have been shown to have strong impacts on personalities, dispositions and behaviour patterns with respect to agents of innovation or enterprise and to reactions of members of the general social environment.

Overall, the greater the degree of social actors' conformity and integration within the normative system and institutional structure of the social environment (social cohesion), and the more isolated a cultural context is from other cultures, the more static and change-resistant society is likely to be. Failure to maintain social integration is itself a seed of social change, as are widening scopes of social reference. Inversely, innovations tend to disturb social equilibrium. Innovators and entrepreneurs are by nature agents of change. Intermingling of cultures proves to engender dynamic environments; even resultant conflicts can give impetus to non-destructive and generally beneficial change. It enhances potential for recognising new opportunities, introducing new factor combinations, and for transgressing culturally imparted, subconscious limits to perception and behaviour. Cultures 'open' to influences from without and conferring a positive connotation to change, innovation, and enterprise tend to flourish.

Change will most likely engender resistance by those with vested interests in the status quo. Stimulus to change never comes from those who benefit or endure no hardship or disadvantages from the status quo. The motivation to change comes from those who recognise an opportunity to profit from change or who are misfits or suffer substantial disadvantages under current circumstances. Since existing oligarchies are by nature opposed to upsetting socio-economic balances whereas innovators and entrepreneurs are also by nature intent on changing conditions to their advantage, conflict models of community power prove to be instrumental for analysis of relationships between new ventures and their social environment. Small, rural communities with multiplex relations between social actors and stable social relationships and identities have been observed to have stronger self-defence mechanisms against 'disruptive' change. Interpersonal relationships in rural communities have been found to emphasise traditional hierarchies, norms, and patterns of cooperation and mutual aid more than competitive behaviour and aims at personal achievement. Another factor of importance is the degree of formalisation of institutions. The greater the degree of formalisation of institutional structures, the greater constraints on innovation and enterprise and the greater will be the effect of protecting and preserving the status quo.

Socio-economic structures rest on a rich foundation of past and present relationships of socio-economic actors characterised by reciprocity, trust, non-opportunistic behaviour, solidarity which all help to minimise 'transaction costs'. In this respect at least, most rural areas enjoy a competitive advantage due to multiplex relationships between social actors.

The socio-cultural environment is a factor making strong imprints on 'organisational cultures' influencing, for instance, goal-setting, modes of communication, role expectations, and personal interaction by members of a firm. It also has been shown to provide, sometimes more, sometimes less forces of push and pull towards new venturing. Even social actors with relatively strong internal 'loci-of-control', such as entrepreneurs, are very much subject to influences from role models, norms, beliefs, and general perception about them. An entrepreneur's relations to regional socio-economic structures depends on his or her status as owner/manager of an existing local enterprise coping with regional sets of opportunities and constraints to survive, as the head of a firm located outside the region making a decision on siting, or as initiator of a new venture.

There are two intersecting groups of primary importance for role models and expectations, for formation and transmission of values and attitudes, for patterns of communication (information source - 'informational support group'), and for material and immaterial support of the innovator or entrepreneur: *Social networks* consist of members of the family house-hold, the extended family, circle of friends and neighbours, colleagues, casual acquaintances, and associates. *Peer groups* consist of individuals of equal rank or standing or possibly occupying a status to which the individual in question aspires. The strong presence of innovative, entrepreneurial role models in these groups increases their visibility for potential agents of innovation and enterprise. Learning more about functions and structures of support groups, about role models of the existing enterprising population, and about perceived opportunities and constraints would be a valuable asset for formulation of regional development strategies. The importance of incubator-environments/learning regions/*milieux innovateurs* for creative synergies, firm spin-offs, and innovation diffusion cannot be stressed enough. The political system and culturally based attitudes are eminent factors in bringing about such environments.

For economic analysis, socio-economic characteristics of the family household have already been recognised as a determinant in the attained level of education and occupational status of individuals. In sociology, psychology, social psychology and cultural anthropology the family house-hold's importance for socialisation of the individual and for personality formation (e.g., through role models, conferred values and attitudes) has repeatedly been a basic premise in investigation of the entrepreneurial personality. In a context of 'overregulation' of the economy, renewed interest can be observed in economic analysis of the family firm's high potential for flexibility in resource acquisition and allocation and for significant savings in 'transaction costs' for provision of financial capital and labour.

Power structures are forceful factors in either facilitating or hindering innovation diffusion and the success of new ventures. Hence, regional power structures cannot be ignored in regional development analysis. Independently-minded, innovative entrepreneurs are, of course, obliged and well disposed to seek alliances and to cooperate with strategic partners in public administration, the finance and business sectors. However, regional development policies should refrain from nuturing, institutionalising and using as agents of distribution of public resources formal 'community elites', 'community networks'. Once firmly established, such structures tend to become the playground of status-conscious, self-enhancing and self-perpetuating power oligarchies frequently interested in maintaining the status-quo and equivocating community welfare with their own. Oligopolistic structures are disposed to discourage or impede new and innovative entries into the regional economy if they themselves cannot profit from the new opportunities created. In comparison, regional economies with diversified structures based on numerous SMEs are, as a rule, more permeable for 'outsiders' or 'newcomers'.

Regional development policies making use of traditional institutional structures (e.g., of the agricultural sector) may initially save transaction costs by utilising existing infrastructural and personal networks. In the long run, however, they tend to reinforce existing, often socially and technologically obsolete structures and to limit their target groups to selected segments of the population. They thus may well put truly innovative newcomers at a disadvantage as compared to favoured interest groups, even with respect to their access to so-called 'public goods'.

Diffusion of new ideas and technological innovations has long been a subject of communications research. Mass media have had a significant impact but are only part of a 'two-step' process in information flow. They are especially instrumental for the 'awareness stage' of information flow. Decision-making still relies on visibility of role models and personal, preferably spontaneous, face-to-face communication. This manifests itself, even in the age of modern electronic communication and travel, in persisting advantages of urban agglomerations, for instance in the location of management headquarters of transnational corporations in 'global cities'. The major advantage of electronic communication and travel media is the expansion of the entrepreneur's 'mean information field' and in increased access to 'weak links' in the periphery of social networks which have been shown to be a potentially rich source of information and strategic support. The continuing significance of 'face to face' relationships, as a major part of the second step in information flow, should not be underestimated.

In conclusion, formulation of adequate policy measures for *sustainable* regional development must be founded on better understanding of non-economic determinants of endogenous innovation and entrepreneurship which are dismissed by conventional, neoclassical economists but which are documented in the foregoing. With respect to the triangular model of resource flows between private households, the state, and the private business sector, we must discard notions of synonymy of entrepreneurship and capital and of solely horizontal, if competitive relationships between leading actors in the business sector. In effect, the SME-owner and manager, the key player in sustainable regional development, now has an intermediate position, depending on customers, employees, the state, 'brokers' and specialised services, as well as on global capital and corporate interests. Gaining and maintaining a share of the global market is important for a dynamic firm's or a new venture's viability and vitality. This depends on the SME-entrepreneur's ability to deal with global corporate and financial management and with resistance from nonreceptive oligarchical environments. Policies for sustainable regional development must aim at strengthening the position of SME-entrepreneurs and potential 'new-venturers' in these relationships, by channelling public resources away from incentive-granting commercial and industrial recruitment which tend to reinforce scale-economies and established structures, by reducing their credit-dependency on speculative credit markets, by reducing regulatory constraints, by assuring exchange and transfer of information and know-how among regional actors (creating local 'incubators' for new businesses, linking local business and educational institutions, sponsoring local competition in science and technology) and, in particular, by providing open public access to modern transportation and communication infra-structure and services providing local enterprises with direct access to global information and technology and to world-wide markets, thus enhancing their ability to participate in transnational production and marketing consortia. This necessitates continually improving regional technical communication and transportation infrastructure and services as well as the education and skills of available regional human capital. However, even with these measures, outlooks for general and sustainable socio-economic equity across space are not promising if national governments do not unite to evolve effective means to regulate speculative global capital markets.

Bibliography

Aberle, D.F. (1961), 'Culture and Socialization', in F.L.K. Hsu (ed), *Psychological Anthropology*, Homewood, Ill., pp. 381ff.

Ajzen, I. (1991), 'The Theory of Planned Behavior', *Organizational Behavior and Human Decision Processes*, vol. 50, pp. 179-211.

Aldrich, H. (1982), 'Origins and Persistence of Social Networks', in P.V. Marsden and N. Lin (eds), *Social Structure and Network Analysis*, Beverly Hills, pp. 281-293.

Aldrich, H. and Zimmer, C. (1986), 'Entrepreneurship Through Social Networks', in D.L. Sexton and R.W. Smilor (eds), *The Art and Science of Entrepreneurship*, Cambridge MA, pp. 3-23.

Aldrich, H.E. (1979), *Organizations and Environments*, Englewood Cliffs.

Alexander, A.P. (1967), 'The Supply of Industrial Entrepreneurship', *Explorations in Entrepreneurial History*, vol. 4, no. 2, pp. 136-149.

Allport, G.W. (1937), *Personality: A Psychological Interpretation*, New York.

Allport, G.W. (1966), 'Traits Revisited', *American Psychologist*, vol. 21, pp. 1-10.

Almond, G.A. and Verba, S. (1965), *The Civic Culture. Political Attitudes and Democracy in Five Nations*, Boston.

Alves, W.R. and Morrill, R.L. (1975), 'Diffusion Theory and Planning', *Economic Geography*, vol. 51, no. 3, pp. 209-304.

Amin, A. and Thrift, N. (eds) (1994), *Globalization, Institutions, and Regional Development in Europe*, Oxford.

Amin, A. and Thrift, N. (1993), 'Globalization, Institutional Thickness and Local Prospects', *Revue d'Economie Régionale et Urbaine*, vol. 3, pp. 405-427.

Arcangeli, F. (1987), 'Verso un Paradigma Economico-spaziale per l'Analisi della Diffusione delle Innovazioni nella Produzione e nel Consumo', in R. Camagni, R. Cappelin e G. Garofoli (eds.), *Cambiamento Technologico e Diffusione Territoriale*, Milano, pp. 87-107.

Arcangeli, F. (1993), 'Local and Global Features of the Learning Process', in Humbert, M. (ed), *The Impact of Globalisation on Europe's Firms and Industries*, London, New York, pp. 34-41.

Asch, S.E. (1951), 'Effects of Group Pressure upon the Modification and Distortion of Judgments', in H. Guetzkow (ed), *Groups, Leadership and Men*, Pittsburgh, pp. 177-190.

Atkinson, J.W. (1957), 'Motivational Determinants of Risk-taking Behavior', *Psychological Review*, vol. 64, no. 6, pp. 359-372.

Aydalot, Ph. (ed.) (1986), *Milieux Innovateurs en Europe*, GREMI-Paris.

Back, K.W. (1951), 'Influence through Social Communication', *Journal of Abnormal and Social Psychology*, vol. 46, pp. 9-23.

Bagnasco, A. (1977), *Tre Italie. La Problematica Territoriale dello Sviluppo Italiano*, Bologna.

Bagnasco, A. (1985), 'La Costruzione Sociale del Mercato: Strategie di Imprese e Esperimenti di Scala in Italia', *Stato e Mercato*, vol. 13.

Bannock, G. (1986), 'The Economic Role of the Small Firm in Contemporary Industrial Society', in J. Curran, J. Stanworth and D. Watkins (eds), The Survival of the Small Firm , vol. I: *The Economics of Survival and Entre-preneurship*, Aldershot, pp. 8-18.

Barnard, C.I. (1949), 'The Entrepreneur and Formal Organization' in Harvard University, Research Center in Entrepreneurial History (ed), *Change and the Environment. Postulates and Patterns for Entrepreneurial History*, Cambridge, MA, pp. 7ff.

Barnes, J.A. (1979), 'Network Analysis: Orienting Notion, Rigorous Technique or Substantive Field of Study?', in P.W. Holland and S. Leinhardt (eds), *Perspectives on Social Network Research*, New York, pp. 403-423.

Barnett, H.G. (1953), *Innovation: The Basis of Cultural Change*, New York

Barreto, H. (1989), *The Entrepreneur in Microeconomic Theory. Disappearance and Explanation*, London.

Bateson, G. (1944), 'Cultural Determinants of Personality', in J.M. Hunt (ed), *Personality and the Behavior Disorders*, vol. 2, New York, pp. 714-735.

Baumol, W.J. (1982), 'Toward Operational Models of Entrepreneurship', in J. Ronen (ed.), *Entrepreneurship*, Lexington MA, pp. 29-48.

Bem, D. and Allen, A. (1974), 'On Predicting Some of the People Some of the Time. The Search for Cross-situational Consistencies in Behavior', *Psychological Review*, vol. 81, no. 6, pp. 506-520.

Benedict, R. (1934), *Patterns of Culture*, New York.

Benedict, R. (1946), 'Study of Cultural Patterns in European Nations', *Transactions of the New York Academy of Science*, Ser. II, vol. 8, pp. 274-279.

Benedict, R. (1953), 'Continuities and Discontinuities in Cultural Conditioning', in C. Kluckhohn and H.A. Murray (eds), *Personality in Nature, Society and Culture*, New York, pp. 522-531.

Bennett, R.J. and McCoshan, A. (1993), *Enterprise and Human Resource Development. Local Capacity Building*, London.

Bergstrom, Th. C. (1996), 'Economics in a Family Way', *Journal of Economic Literature*, vol. 34, December, pp. 1903-1934.

Bhide, A. (1994), 'How Entrepreneurs Craft Strategies That Work', *Harvard Business Review*, nos. 3 and 4, pp. 150-161.

Bierstedt, R. (1970), *The Social Order*, New York.

Binks, M. and Jennings, A. (1986), 'Small Firms as a Source of Economic Rejuvenation', in J. Curran, J. Stanworth and D. Watkins (eds), The Survival of the Small Firm, vol. I: *The Economics of Survival and Entrepreneurship*, Aldershot, pp. 19-38.

Birley, S. (1985), 'The Role of Networks in the Entrepreneurial Process', *Journal of Business Venturing*, no. 1, pp. 107-117.

Bishop, G.F. (1968), 'Context Effects on Self-perception of Interest in Government and Public Affairs', in H.-J. Hippler, N. Schwarz and S. Sudman (eds), *Social Information Processing and Survey Methodology*, New York, pp. 179ff.

Black, S., Bryden, J. and Sproull, A. (1996), 'Telematics, Rural Economic Development and SMEs: Some Demand-side Evidence', *Informationen zur Raumentwicklung*, nos. 11 and 12, pp. 755-775.

Blaikie, P. (1978), 'The Theory of the Spatial Diffusion of Innovations: A Spacious Cul-de-sac', *Progress in Human Geography*, vol. 2, no. 2, pp. 268-295.

Blaut, J.M. (1977), 'Two Views of Diffusion', *Annals of the Association of American Geographers*, vol. 67, no. 3, pp. 343-349.

Bodenhausen, G.V. and Wyer, R.S. (1968), 'Social Cognition and Social Reality: Information Acquisition and Use in the Laboratory and the Real World', in H.-J. Hippler, N. Schwarz and S. Sudman (eds), *Social Information Processing and Survey Methodology*, New York, pp. 6ff.

Bögenhold, D. and Staber, U. (1991), 'The Decline and Rise of Self-employment', *Work, Employment and Society*, vol. 5, no. 2, pp. 223-239.

Brittain, J.W. and Freeman, J.H. (1981), 'Organizational Proliferation and Density Dependent Selection', in: J.R. Kimberly and R.H. Miles (eds), *The Organizational Life Cycle*, San Francisco, pp. 291ff.

Brockhaus, R.H. (1980), 'Risk Taking Propensity of Entrepreneurs', *Academy of Management Journal*, vol. 23, no. 3, pp. 509-520.

Brockhaus, R.H. (1982), 'The Psychology of the Entrepreneur', in C.A. Kent, D.L. Sexton and K.H. Vesper (eds), *Encyclopedia of Entrepreneurship*, Englewood Cliffs NJ, pp. 39-71.

Brockhaus, R.H. and Horwitz, P.S. (1986), 'The Psychology of the Entrepreneur', in D.L. Sexton and R.W. Smilor (eds), *The Art and Science of Entrepreneurship*, Cambridge MA, pp. 25-47.

Brockhaus, R.H. and Nord, W.R. (1979), 'An Exploration of Factors Affecting the Entrepreneurial Decision: Personal Characteristics vs. Environmental Conditions', *Academy of Management Proceedings*, pp. 364-368.

Brown, L.A. (1975), 'The Market and Infrastructure Context of Adoption: A Spatial Perspective on the Diffusion of Innovation', *Economic Geography*, vol. 51, vol. 3, pp. 185-216.

Brown, L.A. (1981), *Innovation Diffusion*, London.

Brown, L.A. and Cox, K.R. (1971), 'Empirical Regularities in the Diffusion of Innovation', *Annals of the Association of American Geographers*, vol. 61, pp. 551-559.

Brown, L.A.; Malecki, E.J. and Spector, A.N. (1976), 'Adopter Categories in a Spatial Context. Alternative Explanations for an Empirical Regularity', *Rural Sociology*, vol. 41, no. 1, pp. 99-118.

Brown, M.A. (1980), 'Attitudes and Social Categories: Complementary Explanations of Innovation-adoption Behavior', *Environment and Planning A*, vol. 12. no. 2, pp. 175-186.

Brown, M.A. (1981), 'Behavioral Approaches to the Geographic Study of Innovation Diffusion: Problems and Prospects', in K.R. Cox and R.G. Golledge (eds), *Behavioral Problems in Geography Revisited*, New York, pp. 123-144.

Bruno, A.V. and Tyebjee, T.T. (1982), 'The Environment for Entrepreneurship', in C.A. Kent, D.L. Sexton and K.H. Vesper (eds), *Encyclopedia of Entrepreneur-ship*, Englewood Cliffs NJ, pp. 288-315.

Brusco, S. (1982), 'The Emilian Model: Productive Decentralisation and Social Integration', *Cambridge Journal of Economics*, no. 6, pp. 167-184.

Bryant, C.R. (1989), 'Entrepreneurs in the Rural Environment', *Journal of Rural Studies*, vol. 5, no. 4, pp. 337-348.

Bryson, L., Finkelstein, L. and Maciver, R.M. (eds) (1947), *Conflicts of Power in Modern Culture*, New York.

Burgess, E.W. and Locke, H.J. (1950), *The Family. From Institution to Compan-ionship*, New York

Burt, R.S. (1982), *Toward a Structural Theory of Action. Network Models of Social Structure, Perception, and Action*, New York, London

Busch, L. (1978), 'On Understanding Understanding: Two Views of Communication', *Rural Sociology*, vol. 43, no. 3, pp. 450-473.

Camagni, R. (1991), 'Introduction: From the Local 'Milieu' to Innovation through Cooperation Networks', in R. Camagni (ed), *Innovation Networks: Spatial Perspectives*, London/New York, pp. 1-9.

Camagni, R. (1991), 'Local "Milieu", Uncertainty and Innovation Networks: Towards a New Dynamic Theory of Economic Space', in R. Camagni (ed), *Innovation Networks: Spatial Perspectives*, London/New York, pp. 121-144.

Camagni, R. (1992), 'Development Scenarios and Policy Guidelines for the Lagging Regions in the 1990's', *Regional Studies*, vol. 26, no. 4, pp. 361-374.

Camagni, R. e Capppelin, R. (1987), 'Cambiamento Strutturale e Dinamica della Produttività nelle Regioni Europee', in R. Camagni, R. Cappelin e G. Garofoli (eds), *Cambiamento Technologico e Diffusione Territoriale*, Milano, pp. 131-269.

Camagni, R.; Cappelin, R.; e Garofoli, G. (eds) (1987), *Cambiamento Technologico e Diffusione Territoriale*, Milano.

Camagni, R.P. (1985), 'Spatial Diffusion of Pervasive Process Innovation', *Papers of the Regional Science Association*, vol. 58, pp. 83-95.

Carsrud, A.L.; Olm, K.W.; and Eddy, G.G. (1986), 'Entrepreneurship. Research in Quest of a Paradigm', in D.L. Sexton and R.W. Smilor (eds), *The Art and Science of Entrepreneurship*, Cambridge MA, pp. 367-378.

Cartwright, D. (1959), 'Power: A Neglected Variable in Social Psychology', in D. Cartwright (ed), *Studies in Social Power*, Ann Arbor, pp. 1ff.

Cartwright, D. (1965), 'Influence, Leadership, Control', in J.G. March (ed), *Handbook of Organizations*, Chicago, pp. 1-47.

Cartwright, D. and Zander, A. (eds) (1968), *Group Dynamics. Research and Theory*, New York

Casson, M. (1982), *The Entrepreneur*, Oxford.

Casson, M. (1991), *The Entrepreneur. An Economic Theory*, Aldershot.

Castel, R. (1996), 'Work and Usefulness to the World', *International Labour Review*, vol. 135, no. 6, pp. 615-622.

Cécora, J. (1991a), *The Role of 'Informal' Activity in Household Economic Behaviour*, vol. 22, Beiträge zur Ökonomie von Haushalt und Verbrauch, Berlin.

Cécora, J. (1991b), *Ressourceneinsatz ländlicher Haushalte für die Lebenshaltung*, vol. 400, Angewandte Wissenschaft, Münster-Hiltrup.

Cécora, J. (1991c), 'Les Échanges entre les Ménages dans l'Ouest de l'Allemagne', *Sociétés Contemporaines*, vol. 8, pp. 43-59.

Cécora, J. (1995), 'Lo Sviluppo del Mondo Rurale: Problemi e Politiche, Istituzioni e Strumenti - la Germania', *Quaderni della Rivista di Economia Agraria*, vol. 20, pp. 449-462.

Cécora, J. (ed) (1993), *Economic Behaviour of Family Households in an International Context. Resource Income and Allocation in Urban and Rural, in Farm and Nonfarm Households*, vol. 295, Society for Agricultural Policy Research and Rural Sociology (FAA), Bonn.

Cécora, J. (ed) (1994), *Changing Values and Attitudes in Family Households with Rural Peer Groups, Social Networks, and Action Spaces. Implications of Institutional Transition in East and West for Value Formation and Transmission*, vol. 296, Society for Agricultural Policy Research and Rural Sociology (FAA), Bonn.

Chanard, A. (1998), *Neue Beschäftigungsformen und Arbeitsplätze im ländlichen Raum*, http-access: www.rural-europe.aeidl.be

Chandler Jr, A.D. (1977), *The Visible Hand: Managerial Revolution in American Business*, Cambridge MA.

Chandler Jr., A.D. (1984), 'Emergence of Managerial Capitalism', *Business History Review*, vol. 58, Winter, pp. 473-503.

Chandler Jr., A.D. (1990), *Scale and Scope: Dynamics of Industrial Capitalism*, Cambridge MA.

Chari, V.V. and Hopenhayn, H. (1991), 'Vintage Human Capital, Growth, and the Diffusion of New Technology', *Journal of Political Economy*, vol. 99, no. 6, pp. 1142-1165.

Chell, E. (1986), 'The Entrepreneurial Personality: A Review and Some Theoretical Developments' in J. Curran, J. Stanworth and D. Watkins (eds), The Survival of the Small Firm, vol. I: *The Economics of Survival and Entrepreneurship*, Aldershot, pp. 102-119.

Chesnais, F. (1993), 'Globalisation, World Oligopoly and Some of Their Implications', in Humbert, M. (ed), *The Impact of Globalisation on Europe's Firms and Industries*, London, New York, pp. 12-21.

Chombart de Lauwe, P. (1959), 'Le Milieu Social et l'Etude Sociologique des Cas Individuals', *Informations Sociales*, vol. 2, pp. 41-54.

Ciciotti, E. (1987), 'Innovazione e Sviluppo Regionale: Alcune Considerazioni sulle Implicazioni di Politica Economica', in R. Camagni, R. Cappelin e G. Garofoli (eds), *Cambiamento Technologico e Diffusione Territoriale*, Milano, pp. 295-315.

Coase, R.H. (1937), 'The Nature of the Firm', *Economica*, vol. 4, pp. 386-405.

Cochran, T.C. (1965), 'The Entrepreneur in Economic Change', *Explorations in Entrepreneurial History*, vol. 3, no. 1, pp. 25-38.

Cole, A.H. (1967), 'An Approach to the Study of Entrepreneurship: A Tribute to Edwin F. Gay', in H.G.J. Aitken (ed), *Explorations in Enterprise*, Cambridge MA, pp. 30-44.

Coleman, J.S. (1988), 'Social Capital in the Creation of Human Capital', *American Journal of Sociology*, vol 94, pp. 95-120.

Colla, L. e Leonardi, G. (1987), 'Modelli di Diffusione dell'Innovazione', in R. Camagni, R. Cappelin e G. Garofoli (eds), *Cambiamento Technologico e Diffusione Territoriale*, Milano, pp. 59-85.

Collins, B.E. and Raven, B.H. (1969), Group Structure: Attraction, Coalitions, Communication, and Power', in G. Lindzey and A. Aronson (eds), Handbook of Social Psychology. (IV): *Group Psychology and Phenomena of Interaction*, Reading MA, pp. 102-204.

Collins, O.F.; Moore, D.G. with Unwalla, D.B. (1964), 'The Enterprising Man', *MSU Business Studies*, East Lansing MI.

Conlisk, J. (1966), 'Why Bounded Rationality?', *Journal of Economic Literature*, vol. 34, no. 2, pp. 669-700.

Cook, K.S. (1982), 'Network Structures from an Exchange Perspective', in P.V. Marsden and N. Lin (eds), *Social Structure and Network Analysis*, Beverly Hills, pp. 177-199.

Cooper, A.C. (1986), 'Entrepreneurship and High Technology', in D.L. Sexton and R.W. Smilor (eds), *The Art and Science of Entrepreneurship*, Cambridge MA, pp. 153-168.

Cooper, A.C. and Dunkelberg, W.C. (1987), 'Entrepreneurial Research: Old Questions, New Answers and Methodological Issues', *American Journal of Small Business*, vol. 11, no. 3, pp. 1-22.

Courlet, C. and Pecqueur, B. (1991), 'Local Industrial Systems and Externalities: An Essay in Typology', *Entrepreneurship and Regional Development*, vol. 3, pp. 305-315.

Crevoisier, O. and Maillat, D. (1991), 'Milieu, Industrial Organization and Territorial Production System: Towards a New Theory of Spatial Development', in R. Camagni (ed), *Innovation Networks: Spatial Perspectives*, London, New York, pp. 13-34.

Cromie, S., Birley, S. and Callaghan, I. (1993), 'Community Brokers: Their Role in the Formation and Development of Business Ventures', *Entrepreneurship and Regional Development*, vol. 5, pp. 247-264.

Dalle, J.-M. et Foray, D. (1995), 'Des Fourmis et des Hommes', *Cahiers d'Economie et Sociologie Rurales*, vol. 37, no. 4, pp. 69-92.

Davidsson, P., Lindmark, L. and Olofsson, C. (1993), 'Regional Characteristics, Business Dynamics and Economic Development', in Karlsson, C., Johannisson, B. and Storey, D. (eds), *Small Business Dynamics. International, National and Regional Perspectives*, London, New York, pp. 145-174.

Davidsson, P. (1995), 'Culture, Structure and Regional Levels of Entrepreneurship', *Entrepreneurship and Regional Development*, vol. 7, pp. 41-62.

De Graaf, P.M. and Huinink, J.J. (1992), 'Trends in Measured and Unmeasured Effects of Family Background on Educational Attainment and Occupational Status in the Federal Republic of Germany', *Social Science Research*, vol. 21, no. 1, pp. 84-112.

De Vos, G. (1961), 'Symbolic Analysis in the Cross-cultural Study of Personality', in B. Kaplan (ed), *Studying Personality Cross-Culturally*, New York, pp. 599-634.

De Vos, G.A. and Hippler, A.A. (1969), 'Cultural Psychology: Comparative Studies of Human Behavior', in G. Lindzey and A. Aronson (eds), Handbook of Social Psychology (IV): *Group Psychology and Phenomena of Interaction*, Reading MA, pp. 323-417.

DIE ZEIT (1997), Various atricles on globalisation, and neo-liberalism by, among others, Uwe Jean Heuser, Antonia Grunenberg, und Bernard Cassen, vol. 12, Sept. 12.

Dmitrieva, E. (1996), 'Orientations, Re-orientations or Disorientations? Expecta-tions of the Future Among Russian School-leavers', in H. Pilkington (ed), *Gender, Generation and Identity in Contemporary Russia*, London, pp. 75-91.

Dollinger, M.J. (1983), 'Use of Budner's Intolerance of Ambiguity Measure for Entrepreneurial Research', *Psychological Reports*, vol. 53, no. 5, pp. 1019-1021.

Doreian, P. and Woodard, K.L. (1992), 'Fixed Versus Snowball Selection of Social Networks', *Social Science Research*, vol. 21, no. 2, pp. 216-233.

Dosi, G. (1998), 'Sources, Procedures and Microeconomic Effects of Innovation', *Journal of Economic Literature*, vol. 26, no. 3, pp. 1120-1171.

Drache, D. and Gertler, M.S. (eds) (1991), *The New Era of Global Competition. State Policy and Market Power*, Montreal, Kingston, London, Buffalo.

Dubini, P. (1988), 'The Influence of Motivations and Environment on Business Start-ups: Some Hints for Public Policies', *Journal of Business Venturing*, vol. 4, pp. 11-26.

Durand, D. and Shea, D. (1974), 'Entrepreneurial Activity as a Function of Achievement Motivation and Reinforcement Control', *Journal of Psychology*, vol. 88, pp. 57-63.

Eisinger, P.K. (1988), *The Rise of the Entrepreneurial State. State and Local Economic Development Policy in the United States*, Madison WI.

EU-LEADER (1998), Presentation of projects and agents of the LEADER-Programme, Manuscript, http-access: www.rural-europe.aeidl.bc

Evans, D.S. and Leighton, L.S. (1989), 'Some Empirical Aspects of Entrepreneurship', *American Economic Review*, vol. 79, no. 3, pp. 519-535.

Executive Forum (1986), 'To Really Learn about Entrepreneurship, Let's Study Habitual Entrepreneurs', *Journal of Business Venturing*, no. 1, pp. 241-243.

Falemo, B. (1989), 'The Firm's External Persons: Entrepreneurs or Network Actors?', *Entrepreneurship and Regional Development*, vol. 1, pp. 167-177.

Feld, S. L. (1981), 'The Focused Organization of Social Ties', *American Journal of Sociology*, vol. 86, no. 5, pp. 1015-1035.

Feldmann, H. (1997), 'Unternehmenskontrolle in Transformationsländern - Die Lehren der Neuen Institutionenökonomik', *Jahrbuch für Wirtschafts-wissenschaften*, vol. 49, no. 3, pp. 288-316.

Festinger, L. (1950), 'Informal Social Communication', *Psychological Review*, vol. 57, pp. 271-282.

Festinger, L. (1957), *A Theory of Cognitive Dissonance*, Stanford.

Festinger, L. (1968), 'Informal Social Communication', in D. Cartwright and A. Zander (eds), *Group Dynamics. Research and Theory*, New York pp. 182-191.

Festinger, L. and Aronson, E. (1968), 'Arousal and Reduction of Dissonance in Social Contexts', in D. Cartwright and A. Zander (eds), *Group Dynamics. Research and Theory*, New York pp. 125-136.

Festinger, L., Schachter, S. and Back, K. (1950), *Social Pressures in Informal Groups*, New York.

Festinger, L., Schachter, S. and Back, K. (1968), 'Operation of Group Standards', in: D. Cartwright and A. Zander (eds), *Group Dynamics. Research and Theory*, New York pp. 152ff.

Flora, J.L., Sharp, J., Flora, C. and Newlon, B. (1997), 'Entrepreneurial Social Infrastructure and Locally Initiated Economic Development in the Nonmetropolitan United States', *The Sociological Quarterly*, vol. 38, no. 4, pp. 623-645.

Foss, O. (1996), 'Structural Change of a Welfare State - Impacts on Rural Areas', *Informationen zur Raumentwicklung*, nos. 11 and 12, pp. 791-807.

Freeman, D.B. (1985), 'The Importance of Being First: Preemption by Early Adopters of Farming Innovations in Kenya', *Annals of the Association of American Geographers*, vol. 75, pp. 17-28.

French Jr., J.R.P. and Raven, B. (1959), 'The Basis of Social Power', in D. Cartwright (ed), *Studies in Social Power*, Ann Arbor, pp. 150-167.

French, Jr., J.R.P. and Snyder, R. (1959), 'Leadership and Interpersonal Power', in D. Cartwright (ed), *Studies in Social Power*, Ann Arbor, pp. 118-149.

Freshwater, D. (1996), 'Globalization and Employment in Rural Economies', *Informationen zur Raumentwicklung*, nos. 11 and 12, pp. 777-790.

Fromm, E. (1941), *Escape from Freedom*, New York.

Fromm, E. (1974), 'Psychoanalytic Characterology and its Application to the Understanding of Culture', in S.S. Sargent and M.W. Smith (eds), *Culture and Personality*, New York, pp. 1-12.

Gabe, T.M. (1996), 'Are Tax Incentives for Economic Development Rational?', *Journal of Regional Analysis and Policy*, vol. 26, no. 1, pp. 99-112.

Gabe, T.M. and Kraybill, D.S. (1998), 'Economic Development Programs and Firm Self-selection', Paper presented at the 37th Annual Meeting of the Southern Regional Science Association in Savannah GA on 3. 4. 1998.

Gallie, D., Gershuny, J. and Vogler, C. (1993), 'Unemployment, the Household, and Social Networks', in D. Gallie, Marsh and C. Vogler (eds), *Social Change and the Experience of Unemployment*, Oxford.

Ganzeboom, H.B.G., De Graaf, P.M., Treiman, D.J. and De Leeuw, J. (1992), 'A Standard International Socio-economic Index of Occupational Status', *Social Science Research*, vol. 21, no. 1, pp. 1-56.

Garofoli, G. (1987), 'Barriere all'Innovazione e Politiche di Intervento a Livello Regionale e Sub-regionale', in R. Camagni, R. Cappelin e G. Garofoli (eds), *Cambiamento Technologico e Diffusione Territoriale*, Milano, pp. 277-294.

Gartner, W.B. (1985), 'A Conceptual Framework for Describing the Phenomenon of New Venture Creation', *Academy of Management Review*, vol. 10, no. 4, pp. 696-706.

Gartner, W.B. (1989), '"Who is an Entrepreneur?" is the Wrong Question', *Entrepreneurship in Theory and Practice*, vol. 13, no. 4, pp. 47-68.

Gartner, W.B. (1989), 'Some Suggestions for Research on Entrepreneurial Traits and Characteristics', *Entrepreneurship Theory and Practice*, vol. 14, no. 4, pp. 27-37.

Gasse, Y. (1986), 'The Development of New Entrepreneurs. A Belief-based Approach', in D.L. Sexton and R.W. Smilor (eds), *The Art and Science of Entrepreneurship*, Cambridge MA, pp. 50ff.

Gatewood, E., Hoy, F. and Spindler, C. (1984), 'Functionalist vs. Conflict Theories: Entrepreneurship Disrupts the Power Structure in a Small Southern Community', in J.A. Hornaday, E.B. Shils, J.A. Timmons, and K.H. Vesper (eds), *Frontiers of Entrepreneurship Research*, Wellesley MA, pp. 265-279.

Gelder, Mr. v. (1997), Draft Opinion of the Committee of the Regions on a Rural Development Policy. http-access: www. rural-europe.aeidl.be

Gerard, H.B. (1954), 'The Effect of Different Dimensions of Disagreement on the Communication Process in Small Groups', *Human Relations*, vol. 6, pp. 249-271.

Gerschenkron, A. (1953), 'Social Attitudes, Entrepreneurship, and Economic Development', *Explorations in Entrepreneurial History*, ser. 1, no. 5, pp. 1-19.

Gershuny, J. (1983), *Social Innovation and the Division of Labour*, Oxford.

Gershuny, J. (1996), 'Occupational Trajectories: The Importance of Longitudinal Evidence in Understanding Change in Social Stratification', in K. de Verhaar, de Goede, van Ophem and de Vries (eds), *On the Challenges of Unemployment in a Regional Europe* (III), Fryske Academy, pp. 167-179.

Gibb, A.A. (1993), 'Key Factors in the Design of Policy Support for the Small and Medium Enterprise (SME) Development Process: An Overview', *Entrepreneurship and Regional Development*, vol. 5, pp. 1-24.

Gibb, C.A. (1969), 'Leadership', in G. Lindzey and A. Aronson (eds), Handbook of Social Psychology. (IV): *Group Psychology and Phenomena of Interaction*, Reading MA, pp. 205ff.

Glade, W.P. (1969), 'Approaches to a Theory of Entrepreneurial Formation', *Explorations in Entrepreneurial History*, vol. 4, no. 3, pp. 245-259.

Glasman, M. (1994), 'The Great Deformation. Polanyi, Poland and the Terrors of Planned Spontaneity', in C.G.A. Bryant and E. Mokrzyeki (eds), *The Great New Transformation?*, London, pp. 191-217.

Gorer, G. (1953), 'Concept of National Character', in C. Kluckhohn and H.A. Murray (eds), *Personality in Nature, Society and Culture*, New York, pp. 246-259.

Gould, P. and Törnquist, G. (1971), 'Information, Innovation and Acceptance (Comments)', in T. Hägerstrand and A.R. Kuklinski (eds),

Information Systems for Regional Development, Lund Studies in Geography, Lund, pp. 148-168.

Granovetter, M. (1974), *Getting a Job. A Study of Contacts and Careers*, Cambridge MA.

Granovetter, M. (1982), 'The Strength of Weak Ties. A Network Theory Revisited', in P.V. Marsden and N. Lin (eds), *Social Structure and Network Analysis*, Beverly Hills, pp. 105-130.

Granovetter, M. (1985), 'Economic Action and Social Structure. The Problem of Embeddedness', *American Journal of Sociology*, vol. 91, no. 3, pp. 481-510.

Greenfield, S.M. and Strickon, A. (1979), 'Entrepreneurship and Social Change: Toward a Populational, Decision-making Approach', in S.M. Greenfield, A. Strickon and R.T. Aubey (eds), *Entrepreneurs in Cultural Context*, Albuquerque NM, pp. 329-351.

Grossman, G.M. and Helpman, E. (1994), 'Endogenous Innovation in the Theory of Growth', *Journal of Economic Perspectives*, vol. 8, no. 1, pp. 23-44.

Hagan, E. (1962), *On the Theory of Social Change: How Economic Growth Begins*, Homewood IL.

Hagen, E.E. (1960), 'The Entrepreneur as a Rebel Against Traditional Society', *Human Organization*, vol. 19, no. 4, pp. 185-187.

Hägerstrand, T. (1965), 'Aspects of the Spatial Structure of Social Communication and the Diffusion of Information', *Papers of the Regional Science Association*, vol. 16, pp. 27-42.

Hägerstrand, T. (1967), *Innovation Diffusion as a Spatial Process*, Chicago.

Hägerstrand, T. (1974), 'On Socio-technical Ecology and the Study of Innovations', *Ethnologica Europaea*, vol. 7, pp. 17-34.

Hall, E.T. (1973), *The Silent Language*, Garden City.

Hall, E.T. (1976), *Beyond Culture*, New York, London.

Hall, E.T. (1982), *The Hidden Dimension*, New York, London.

Hall, E.T. (1983), *The Dance of Life. The Other Dimension of Time*, New York.

Hanham, R.Q. and Brown, L.A. (1976), 'Diffusion Waves within the Context of Regional Economic Development' *Journal of Regional Science*, vol. 16, no. 1, pp. 65-71.

Hansch, E. and Piorkowsky, M.-B. (1997), 'Haushalts-Unternehmens-Komplexe: Untersuchungsgegenstand, Forschungsprogramm, haushalts-ökonomische Perspektiven', *Hauswirtschaft und Wissenschaft*, no. 1, pp. 3-10.

Hansen, N. (1992), 'Competition, Trust, and Reciprocity in the Development of Innovative Regional Milieux', *Papers in Regional Science*, vol 71, no. 2, pp. 95-105.

Hassinger, E. (1959), 'Stages in the Adoption Process', *Rural Sociology*, vol. 24, pp. 52-53.

Hayek, F.A. von (1937), 'Economics and Knowledge', *Economica*, vol. 4, pp. 33-54.

Heider, F. (1946), 'Attitudes and Cognitive Organization', *Journal of Psychology*, vol. 21, pp. 107-112.

Hemphill, J.K. (1961), 'Why People Attempt to Lead', in L. Petrullo and B. Bass (eds), *Leadership and Interpersonal Behavior*, New York, pp. 201-215.

Higgins, B. and Savoie, D.J. (1995), *Regional Development Theories and Their Application*, New Brunswick.

Hisrich, R.D. (1986), 'The Woman Entrepreneur. Characteristics, Skills, Problems, and Prescriptions for Success', in D.L. Sexton and R.W. Smilor (eds), *The Art and Science of Entrepreneurship*, Cambridge MA, pp. 61-81.

Hollander, E.P. (1961), 'Some Effects of Perceived Status on Responses to Innovative Behavior', *Journal of Abnormal and Social Psychology*, vol. 63, no. 2, pp. 247-250.

Honigmann, J.J. (1973 (1954)), *Culture and Personality*, Westport CT.

Hornaday, J.A. and Aboud, J. (1971), 'Characters of Successful Entrepreneurs', *Personnel Psychology*, vol. 24, pp. 141-153.

Hornaday, J.A. and Bunker, Ch.S. (1970), 'The Nature of the Entrepreneur', *Personnel Psychology*, vol. 23, pp. 47-54.

Hoy, F. and Vaught, B.C. (1980), 'The Rural Entrepreneur. A Study in Frustration', *Journal of Small Business Management*, vol. 18, pp. 19-24.

Hsu, F.L.K. (1961), 'Kinship and Ways of Life: An Exploration', in F.L.K. Hsu (ed), *Psychological Anthropology*, Homewood IL, pp. 400ff.

Hull, D.L., Bosley, J.J. and Udell, G.G. (1980), 'Renewing the Hunt for the Heffalump: Identifying Potential Entrepreneurs by Personality Characteristics', *Journal of Small Business Management*, vol. 18, pp. 11-18.

Humbert, M. (1993), 'Introduction: Questions, Constraints and Challenges in the Name of Globalisation', in Humbert, M. (ed), *The Impact of Globalisation on Europe's Firms and Industries*, London, New York, pp. 3-11.

Hymes, D.H. (1961), 'Linguistic Aspects of Cross-cultural Personality Study', in B. Kaplan (ed), *Studying Personality Cross-Culturally*, New York, pp. 313-359.

Inkeles, A. (1953), 'Some Sociological Observations on Culture and Personality Studies', in C. Kluckhohn and H.A. Murray (eds), *Personality in Nature, Society and Culture*, New York, pp. 577-592.

Inkeles, A. and Levinson, D.J. (1969), 'National Character: The Study of Modal Personality and Sociocultural Systems', in G. Lindzey and A. Aronson (eds), Handbook of Social Psychology. (IV): *Group Psychology and Phenomena of Interaction*, Reading MA, pp. 418-506.

Irmen, E. und Blach, A. (1996), 'Typen ländlicher Entwicklung in Deutschland und Europa', *Informationen zur Raumentwicklung*, nos. 11 and 12, pp. 713-728.

Jacobs, R.C. and Campbell, D.T. (1961), 'Perpetuation of an Arbitrary Tradition Through Several Generations of a Laboratory Microculture', *Journal of Abnormal and Social Psychology*, vol. 62, pp. 649-658.

Jarillo, J.C. (1989), 'Entrepreneurship and Growth: The Strategic Use of External Resources', *Journal of Business Venturing*, vol. 4, pp. 133-147.

Jenks, L.H. (1949), 'Role Structure of Entrepreneurial Personality', in Harvard University, Research Center in Entrepreneurial History (ed), *Change and the Environment. Postulates and Patterns for Entrepreneurial History*, Cambridge MA, pp. 108-153.

Johannisson, B., Alexanderson, O., Nowicki, K. and Senneseth, K. (1994), 'Beyond Anarchy and Organization: Entrepreneurs in Contextual Networks', *Entrepreneurship and Regional Development*, vol. 6, pp. 329-356.

Juster, F.T. (1990), 'Rethinking Utility Theory', *Journal of Behavioral Economics*, vol. 19, no. 2, pp. 155-179.

Kamann, D.-J. and Strijker, D. (1991), 'The Network Approach: Concepts and Applications', in R. Camagni (ed), *Innovation Networks: Spatial Perspectives*, London, New York, pp. 145-173.

Kanbur, S.M. (1979), 'Of Risk Taking and the Personal Distribution of Income', *Journal of Political Economy*, vol. 87, no. 4, pp. 760-797.

Kaplan, B. (1961), 'Personality Study and Culture', in B. Kaplan (ed), *Studying Personality Cross-Culturally*, New York, pp. 301-311.

Katz, D. (1960), 'The Functional Approach to the Study of Attitudes', *Public Opinion Quarterly*, vol. 24, pp. 163-204.

Katz, E. (1960), 'Communication Research and the Image of Society. Convergence of Two Traditions', *American Journal of Sociology*, vol. 65, no. 5, pp. 435-440.

Katz, E.; Hamilton, H. and Levin, M.L. (1963), 'Traditions of Research on the Diffusion of Innovation', *American Sociological Review*, vol. 28, pp. 237-252.

Katz, M.L. and Shapiro, C. (1985), 'Network Externalities, Competition, and Compatibility', *American Economic Review*, vol. 75, no. 3, pp. 424-440.

Keeble, D and Wever, E. (1986), 'Introduction', in D. Keeble and E. Wever (eds), *New Firms and Regional Development in Europe*, London, Sydney, pp. 1-34.

Kegerreis, R.J., Engel, J.F. and Blackwell, R.D. (1970), 'Innovativeness and Diffusiveness: A Marketing View of the Characteristics of Earliest Adopters', in Kollat, Blackwell and Engel (eds), *Research in Consumer Behavior*, New York.

Kelman, H.C. (1961), 'Processes of Opinion Change', *Public Opinion Quarterly*, vol. 25, pp. 57-78.

Kent, C.A. (1982), 'Entrepreneurship in Economic Development', in C.A. Kent, D.L. Sexton and K.H. Vesper (eds), *Encyclopedia of Entrepreneurship*, Englewood Cliffs NJ, pp. 237-256.

Kets deVries, M.F.R. (1985), 'Dark Side of Entrepreneurship', *Harvard Business Review*, vol. 85, no. 6, pp. 160-167.

Kets deVries, M.F.R. and Miller, D. (1986), 'Personality, Culture, and Organization', *Academy of Management Review*, vol. 11, no. 2, pp. 266-279.

Kierulff, H.E. (1975), 'Can Entrepreneurs Be Developed?', *MSU Business Topics*, vol. 23, pp. 39-44.

Kihlstrom, R.E. and Laffont, J.-J. (1979), 'A General Equilibrium Entrepreneurial Theory of Firm Formation Based on Risk Aversion', *Journal of Political Economy*, vol. 87, no. 4, pp. 719-748.

Kilkenny, M., Nalbarte, L., and Besser, T. (1998), 'Reciprocated Community Support and Small Town - Small Business Success', Contribution to the 37th Meeting of the Southern Regional Science Association, Savannah GA, 1998

Kirzner, I.M. (1974), *Competition and Entrepreneurship*, Chicago.

Kirzner, I.M. (1982), 'The Theory of Entrepreneurship in Economic Growth', in C.A. Kent, D.L. Sexton and K.H. Vesper (eds), *Encyclopedia of Entrepreneur-ship*, Englewood Cliffs NJ, pp. 272-276.

Kirzner, I.M. (1997), 'Entrepreneurial Discovery and the Competitive Market Process: An Austrian Approach', *Journal of Economic Literature*, vol. 35, no. 1, pp. 60-85.

Kluckhohn, C. and Murray, H.A. (1953), 'Personality Formation: The Determin-ants', in: C. Kluckhohn and H.A. Murray (eds), *Personality in Nature, Society and Culture*, New York, pp. 53-67.

Knight, F.H. (1921), *Risk, Uncertainty and Profit*, Boston.

Komarovsky, M. (1961), 'Class Differences in Family Decision-making on Expenditures', in N.N. Foote (ed), *Household Decision-Making*, New York, pp. 255-265.

Komives, J.L. (1972), 'A Preliminary Study of the Personal Values of High Tech-nology Entrepreneurs', in A.C. Cooper and J.L. Komives (eds), *Technical Entre-preneurship*, Symposium-Milwaukee Center for Venture Management, pp. 231-242.

Koppel, B.M. (ed.) (1995), *Induced Innovation Theory and International Agri-cultural Development. A Reassessment*, Baltimore.

Kourilsky, M. (1980), 'Predictors of Entrepreneurship in a Simulated Economy', *Journal of Creative Behavior*, vol. 14, no. 3, pp. 175-198.

Kreps, D. (1990), 'Corporate Culture and Economic Theory', in J. Alt and K. Shepsle (eds), *Perspectives in Political Economy*, New York, pp. 90-143.

Kroeber, A.L. and Parsons, T. (1958), 'Concepts of Culture and of Social System', *American Sociological Review*, vol. 23, no. 10, pp. 582-583.

Krueger, N.F. and Carsrud, A.L. (1993), 'Entrepreneurial Intentions: Applying the Theory of Planned Behaviour', *Entrepreneurship and Regional Development*, vol. 5, pp. 315-330.

Lachman, R. (1980), 'Toward Measurement of Entrepreneurial Tendencies', *Management International Review*, vol. 20, no. 2, pp. 108-116.

Lamont, L.M. (1972), 'What Entrepreneurs Learn from Experience', *Journal of Small Business Management*, vol. 10, pp. 36-41.

Lash, S. and Urry, J. (1987), *The End of Organized Capitalism*, Cambridge.

Lazonick, W. (1991), *Business Organization and the Myth of the Market Economy*, Cambridge MA.

Le Guidec, R. (1996), 'Decline and Resurgence of Unremunerated Work', *Inter-national Labour Review*, vol. 135, no. 6, pp. 645-651.

Lee, E. (1997), 'Globalization and Labour Standards: A Review of Issues', *Inter-national Labour Review*, vol. 136, no. 2, pp. 173-189.

Leibenstein, H. (1976), *Beyond Economic Man. A New Foundation for Micro-economics*, Cambridge MA.

Lerner, D. (1958), *The Passing of Traditional Society. Modernizing the Middle East*, Glencoe IL.

LeRoy, S.F. and Singell, L.D. (1987), 'Knight on Risk and Uncertainty', *Journal of Political Economy*, vol. 95, no. 2, pp. 394-407.

Lewin, S.B. (1996), 'Economics and Psychology: Lessons for Our Own Day from the Early Twentieth Century', *Journal of Economic Literature*, vol. 34, no. 3, pp. 1293-1323.

Lieberman, S. (1956), 'The Effects of Changes in Roles on the Attitudes of Role Occupants', *Human Relations*, vol. 9, pp. 385-402.

Lindesmith, A.R. and Strauss, A.L. (1950), 'A Critique of Culture-personality Writings', *American Sociological Review*, vol. 15, no. 5, pp. 587-600.

Linton, R. (1974), 'Problems of Status Personality', in: S.S. Sargent and M.W. Smith (eds), *Culture and Personality*, New York, pp. 163-173.

Linton, R. (1981 [1945]): *The Cultural Background of Personality*, Westport CT.

Lorenzi, F. (1996), 'Des Stratégies de Développement pour les Zones Rurales de l'Union Européenne', *Informationen zur Raumentwicklung*, nos. 11 and 12, pp. 809-821.

Low, M.B. and MacMillan, I.C. (1988), 'Entrepreneurship: Past Research and Future Challenges', *Journal of Management*, vol. 14, no. 2, pp. 139-162.

Loy, B.A. and Loveridge, S. (1998), 'Willingness to Pay for Industrial Recruitment: a Look at Local Decision-maker Activity', Paper presented at the 37th Annual Meeting of the Southern Regional Science Association in Savannah GA on 3. 4. 1998.

Lucas, R.E. (1988), 'On the Mechanics of Economic Development', *Journal of Monetary Economics*, vol. 22, pp. 3-42.

MacDonald, R. (1996), 'Welfare Dependency, the Enterprise Culture and Self-employed Survival', *Work, Employment and Society*, vol. 10, no. 3, pp. 431-447.

Macdonald, S. and Williams, C. (1994), 'The Survival of the Gatekeeper', *Research Policy*, vol. 23, pp. 123-132.

MacMillan, I.C. (1983), 'The Politics of New Venture Management', *Harvard Business Review*, vol. 61, no. 6, pp. 8-16.

Maillat, D. (1995), 'Territorial Dynamics, Innovative Milieus and Regional Policy', *Entrepreneurship and Regional Development*, vol. 7, pp. 157-165.

Malecki, E.J. (1994), 'Entrepreneurship in Regional and Local Development', *International Regional Science Review*, vol. 16, nos. 1 and 2, pp. 119-153.

Malecki, E.J. (1998), 'How Development Occurs: Local Knowledge, Social Capital, and Institutional Embeddedness', Paper presented at the 37th Annual Meeting of the Southern Regional Science Association in Savannah GA on April 2, 1998.

Malecki, E.J. and Tootle, D.M. (1996), 'The Role of Networks in Small Firm Competitiveness', *International Journal of Technology Management*, vol. 1, nos. 1 and 2, pp. 43-57.

Mann, R.D. (1959), 'A Review of the Relationships Between Personality and Per-formance in Small Groups', *Psychological Bulletin*, vol. 56, no. 4, pp. 241-270.

Marble, D.F. and Nystuen, J.D. (1963), 'An Approach to the Direct Measurement of Community Mean Information Fields', *Papers and Proceedings of the Regional Science Association*, vol. 11, pp. 99-109.

March, J.G. (1955), 'An Introduction to the Theory and Measurement of Influence', *American Political Science Review*, vol. 49, pp. 431-451.

Marchesnay, M. and Julien, P.-A. (1990), 'The Small Business: As a Transaction Space', *Entrepreneurship and Regional Development*, vol. 2, pp. 267-277.

Markusen, A.R. (1985), *Profit Cycles, Oligopoly, and Regional Development*, Cambridge MA, London.

Marlowe, D. and Gergen, K.J. (1969), 'Personality and Social Interaction', in G. Lindzey and A. Aronson (eds), Handbook of Social Psychology. (III): *The Individual in a Social Context*, Reading MA, pp. 590-665.

Marsden, P.V. (1983), 'Restricted Access in Networks and Models of Power', *American Journal of Sociology*, vol. 88, no. 4, pp. 686-717.

Martin, M.J.C. (1984), *Managing Technological Innovation and Entrepreneurship*, Reston VA.

McAllister, L. and Fischer, C.S. (1978), 'A Procedure for Surveying Personal Networks', *Sociological Methods and Research*, vol. 7, no. 2, pp. 131-148.

McClelland, D.C. (1962), 'Business Drive and National Achievement', *Harvard Business Review*, no. 4, pp. 110ff.

McClelland, D.C. (1976), *The Achieving Society*, New York.

McClelland, D.C., Atkinson, J.W., Clark, R.A. and Lowell, E.L. (1953), *The Achievement Motive*, New York.

McGaffey, T.N. and Christy, R. (1975), 'Information Processing Capability as a Predictor of Entrepreneurial Effectiveness', *Academy of Management Journal* (Research Notes), vol. 18, no. 4, pp. 857-863.

McGuire, W.J. (1969), 'The Nature of Attitudes and Attitude Change' in G. Lindzey and A. Aronson (eds), Handbook of Social Psychology. (III): *The Individual in a Social Context*, Reading MA, pp. 136-314.

Meager, N. (1992), 'Does Unemployment Lead to Self-employment?' *Small Business Economics*, vol. 4, pp. 87-103.

Meager, N. (1992), 'The Fall and Rise of Self-employment (again). A Comment on Bögenhold and Staber', *Work, Employment and Society*, vol. 6, no. 1, pp. 127-134.

Méda, D. (1996), 'New Perspectives on Work as Value', *International Labour Review*, vol. 135, no. 6, pp. 633-643.

Merlo, M. and Manente, M. (1994), 'Consequences of Common Agricultural Policy for Rural Development and the Environment', in EUROPEAN COMMISSION, DG-Economic and Financial Affairs (ed), *European Economy Reports and Studies*, vol. 5, pp. 133-164.

Merton, R. (1957), *Social Theory and Social Structure*, Glencoe IL.

Merton, R.K. (1957), The Role-set: Problems in Sociological Theory', *British Journal of Sociology*, vol. 8, pp. 106-120.

Messinger S.L. and Clark, B.R. (1961), 'Individual Character and Social Constraint: A Critique of David Riesman's Theory of Social Conduct', in S.M. Lipset and L. Lowenthal (eds), *Culture and Social Character*, New York, pp. 72-85.

Meyer, H. v. (1996), 'OECD-Indikatoren zur ländlichen Entwicklung', *Informationen zur Raumentwicklung*, nos. 11and 12, pp. 729-743.

Michalos, A. (1997), *Good Taxes: The Case for Taxing Foreign Currency Exchange and Other Financial Transactions*, Toronto.

Mikesell, M.W. (1978), 'Tradition and Innovation in Cultural Geography', *Annals of the Association of American Geographers*, vol. 68, no. 1, pp. 1-16.

Milgram, S. (1961), 'Nationality and Conformity', *Scientific American*, vol. 205, no. 6, pp. 45-51.

Milgrom, P. and Roberts, J. (1990), 'Bargaining Costs, Influence Costs, and the Organization of Economic Activity', in J. Alt and K. Shepsle (eds), *Perspectives in Political Economy*, New York, pp. 57-89.

Miller, D.R. (1961), 'Personality and Social Interaction', in B. Kaplan (ed), *Studying Personality Cross-Culturally*, New York, pp. 271-298.

Miller, G.J. (1992), *Managerial Dilemnas. The Political Economy of Hierarchy*, Cambridge.

Mingione, E. (1991), *Fragmented Societies*, Cambridge.

Mingione, E. e Pugliese, E. (1988), 'La Questione Urbana e Rurale: Tra Super-amento Teorico e Problemi di Confini Incerti', *La Critica Sociologica*, vol. 85, pp. 17-50.

Mischel, W. (1968), *Personality and Assessment*, New York.

Mischel, W. (1973), 'Toward a Cognitive Social Learning Reconceptualization of Personality', *Psychological Review*, vol. 80, no. 4, pp. 252-283.

Mises, L. von (1949), *Human Action. A Treatise on Economics*, Chicago.

Molle, W.T.M. (1987), 'Potenziali Regionali di Innovazione nella Comunità Europea', in R. Camagni, R. Cappelin e G. Garofoli (eds), *Cambiamento Technologico e Diffusione Territoriale*, Milano, pp. 109-127.

Momigliano, F. (1987), 'Revisione di Modelli Interpretativi delle Determinanti ed Effetti dell'Attività Innovativa, della Aggregazione Spaziale dei Centri di R&S e della Diffusione Intraindustriale e Territoriale delle Innovazioni Technologiche', in R. Camagni, R. Cappelin e G. Garofoli (eds), *Cambiamento Technologico e Diffusione Territoriale*, Milano, pp. 19-57.

Monsted, M. (1993), 'Regional Network Processes: Networks for the Service Sector or Development of Entrepreneurs?', in Karlsson, C., Johannisson, B. and Storey, D. (eds), *Small Business Dynamics. International, National and Regional Perspectives*, London, New York, pp. 204-222.

Morrill, R.L. (1970), 'The Shape of Diffusion in Space and Time', *Economic Geography*, vol. 46, pp. 259-268.

Morrill, R.L. and Manninen, D. (1975), 'Critical Parameters of Spatial Diffusion Processes', *Economic Geography*, vol. 51, no. 3, pp. 269-277.

Neander, E. und Schrader, H. (1996), 'Regionale Wirtschaftsförderung in ländlichen Räumen', in K. Eberstein (ed), *Handbuch der regionalen Wirtschaftsförderung*, (Chap. VII), Köln.

Newcomb, T.M. (1953), 'An Approach to the Study of Communicative Acts', *Psychological Review*, vol. 60, no. 6, pp. 393-404.

Nittykangas, H. (1966), 'Entrepreneurship in Rural Areas', English supplement to *Maaseudun uusi aika*, pp. 53-65.

O'Brien, R. (1992), *Global Financial Integration: The End of Geography*, London.

Oksa, J. and Rannikko, P. (1966), 'The Changing Meanings of Rurality Challenge Rural Policies', English supplement to *Maaseudun uusi aika*.

Oliver, C. (1996), 'The Institutional Embeddedness of Economic Activity', in Baum, J.A.C. and Dutton, J.E. (eds), *Advances in Strategic Management*, vol. 13, pp. 163-186.

Orsenigo, L. (1995), 'Technological Regimes, Patterns of Innovative Activities and Industrial Dynamics', *Cahiers d'Economie et Sociologie Rurales*, vol. 37, no. 4, pp. 25-67.

Owens, R.L. (1978), 'The Anthropological Study of Entrepreneurship', *Eastern Anthropologist*, vol. 31, no. 1, pp. 65-80.

Paci, R. (1997), 'More Similar and Less Equal: Economic Growth in the European Regions', *Weltwirtschaftliches Archiv*, vol. 133, no. 4, pp. 609-634.

Pack, H. (1994), 'Endogenous Growth Theory: Intellectual Appeal and Empirical Shortcomings', *Journal of Economic Perspectives*, vol. 8, no. 1, pp. 55-72.

Palmer, M. (1971), 'The Application of Psychological Testing to Entrepreneurial Potential', *California Management Review*, vol. 13, no. 3, pp. 32-39.

Pandey, J. and Tewary, N.B. (1979), 'Locus of Control and Achievement Values of Entrepreneurs', *Journal of Occupational Psychology*, vol. 52, pp. 107-111.

Paqué, K.-H. (1995), 'Technologie, Wissen und Wirtschaftspolitik - Zur Rolle des Staates in Theorien des endogenen Wachstums', *Die Weltwirtschaft*, no. 3, pp. 237-253.

Parsons, T. and White, W. (1961), 'The Link Between Character and Society', in S.M. Lipset and L. Lowenthal (ed), *Culture and Social Character*, New York, pp. 89-135.

Pemberton, H.E. (1937), 'Culture-diffusion Gradients', *American Journal of Sociology*, vol. 42, pp. 226-233.

Pennings, J.M. (1981), 'Environmental Influences on the Creation Process', in J.R. Kimberly and R.H. Miles (eds), *The Organizational Life Cycle*, San Francisco, pp. 135-160.

Perrin, J.-C. (1991), 'Technological Innovation and Territorial Development: An Approach in Terms of Networks and Milieux', in R. Camagni (ed), *Innovation Networks: Spatial Perspectives*, London, New York, pp. 35-54.

Polanyi, K. (1944), *The Great Transformation*, Boston.

Polanyi, K. (1957), 'The Economy as Instituted Process', in K. Polanyi, C.M. Arensberg, and H.W. Pearson (eds), *Trade and Market in the 'Early Empires'. Economies in History and Theory*, New York, London, pp. 243-270.

Polanyi, K. (1977), *The Livelihood of Man*, New York.

Pollak, R.A. (1985), 'A Transaction Cost Approach to Families and Households', *Journal of Economic Literature*, no. 2, pp. 581-608.

Praag, M. Van and Ophem, H. Van (1995), 'Determinants of Willingness and Opportunity to Start as an Entrepreneur', *Kyklos*, vol. 48, no. 4, pp. 513-540.

Quah, D.T. (1996), 'Regional Convergence Clusters Across Europe', Discussion Paper Nr. 1286 of the Centre for Economic Policy Research, London.

Quévit, M. (1991), 'Innovative Environments and Local/International Linkages in Enterprise Strategy: A Framework for Analysis', in R.

Camagni (ed), *Innovation Networks: Spatial Perspectives*, London, New York, pp. 55-70.

Quick, M. (1994), 'Die Vergleichbarkeit territorialer Einheiten in der komparativen Europaforschung', *Europa Regional*, vol. 2, no. 3, pp. 20-29.

Radermacher, F.J. (1997), 'Globalisierung, nachhaltige Entwicklung und Zukunft der Arbeit', Vortrag zur Messeröffnung 'Harz und Heide' Braunschweig, May 3, 1997.

Rainey, D.V. (1998), 'Economic Development Incentives and Capital Location', Paper presented at the 37th Annual Meeting of the Southern Regional Science Association in Savannah GA on April 3, 1998.

Rapoport, A. (1953), 'Spread of Information Through a Population with Socio-structural Bias (I): Asumption of Transitivity', *Bulletin of Mathematical Bio-physics*, vol. 15, pp. 523-533.

Ratti, R. (1991), 'Small and Medium-size Enterprises, Local Synergies and Spatial Cycles of Innovation', in R. Camagni (ed), *Innovation Networks: Spatial Perspectives*, London, New York, pp. 71-88.

Reich, R.B. (1991), *The Work of Nations. Preparing Ourselves for 21st-Century Capitalism*, New York.

Reynolds, P.; Storey, D.J.; and Westhead, P. (1994), 'Cross-national Comparisons of the Variation in New Firm Formation Rates', *Regional Studies*, vol. 28, no. 4, pp. 443-456.

Reynolds, P.D. (1992), 'Predicting New-Firm Interactions of Organisational and Human Populations', in D.L. Sexton and J.D. Kasarda (eds), *The State of the Art of Entrepreneurship*, Boston, pp. 268-297.

Riesman, D. (1950), *The Lonely Crowd*, New Haven 1950, also cited in Parsons and White in Lipset and Lowenthal 1961, p. 89ff.

Rippl, S. (1995), 'Netzwerkanalyse und Intergruppenkontakte: Die persönlichen Beziehungen zwischen Ost- und Westdeutschen', *ZUMA-Nachrichten*, vol. 19, no. 37, November, pp. 76-101.

Roberts, K. and Jung, B. (1995), *Poland's First Post-Communist Generation*, Avebury.

Roberts, K., Adibekian, A., Nemiria, G. and Tarkhnishvili, L. (1997), 'The Young Self-Employed in Armenia, Georgia and Ukraine', Manuscript presented at the 33rd World Congress of the International Institute of Sociology, Köln.

Roberts, K., Clark, S.C. and Wallace, C. (1994), 'Flexibility and Individualisation: A Comparison of Transition into Employment in England and Germany', *Sociology*, vol. 28, no. 1, pp. 31-54.

Robertson T.S. (1971), *Innovative Behavior and Communication*, New York.

Rodwin, L. (1991), 'European Industrial Change and Regional Economic Transformation: An Overview of Recent Experience', in L. Rodwin and

H. Sazanami (eds), *Industrial Change and Regional Economic Transformation*, London, pp. 3-38.

Rogers, E.M. and Shoemaker, F.F. (1971), *Communication of Innovations. A Cross-Cultural Approach*, New York.

Romer, P.M. (1986), 'Increasing Returns and Long-run Growth', *Journal of Political Economy*, vol. 94, no. 5, pp. 1002-1037.

Romer, P.M. (1994), 'The Origins of Endogenous Growth', *Journal of Economic Perspectives*, vol. 8, no. 1, 3-22.

Ronstadt, R. (1984), *Entrepreneurship*, Dover MA.

Ronstadt, R. et al. (eds) (1986), *Frontiers of Entrepreneurship Research*, Welesley MS.

Rosenberg, M.J. (1960), 'A Structural Theory of Attitude Dynamics', *Public Opinion Quarterly*, vol. 24, pp. 319-340.

Rosenberg, N. (1969), 'The Direction of Technological Change: Inducement Mechanisms and Focusing Devices', *Economic Development and Change*, vol. 18, pp. 1-24.

Rotter, J.B. (1966), 'Generalized Expectancies for Internal Versus External Control of Reinforcement', *Psychological Monographs: General and Applied*, vol. 80, no. 1 (or whole no. 609), pp. 1-28.

Ruttan, V.W. (1996), 'What Happened to Technology Adoption-diffusion Research?', *Sociologia Ruralis*, vol. 36, no. 1, pp. 51-73.

Ruttan, V.W. and Hayami, Y. (1984), 'Toward a Theory of Induced Institutional Innovation', *Journal of Development Studies*, vol. 20, no. 4, pp. 204-224.

Sandberg, W.R. and Hofer, Ch.W. (1987), 'Improving New Venture Performance: The Role of Strategy, Industry Structure, and the Entrepreneur', *Journal of Business Venturing*, no. 2, pp. 5-28.

Sassen, S. (1994), *Cities in a World Economy*, Thousand Oaks.

Sassen, S. (1996), 'Whose City Is It? Globalization and the Formation of New Claims', *Public Culture*, vol. 8, no. 2, pp. 205-223.

Sassen, S. (1996), *Losing Control?*, New York.

Schein, E.H. (1960), 'Interpersonal Communication, Group Solidarity, and Social Influence', *Sociometry*, vol. 23, pp. 148-161.

Schein, E.H. (1983), 'The Role of the Founder in Creating Organizational Culture', *Organizational Dynamics*, vol. 12, no. 1, pp. 13-28.

Schere, J.L. (1982), 'Tolerance of Ambiguity as a Discriminating Variable Between Entrepreneurs and Managers', *Academy of Management Proceedings*, New York, pp. 404-408.

Schmidt, H. (1997), 'Der Paragraphenwust tötet den Unternehmergeist', *Die ZEIT*, vol. 15, April 4, 1997, p. 3.

Schrader, H. (1994), 'Impact Assessment of the EU Structural Funds to Support Regional Economic Development in Rural Areas of Germany', *Journal of Rural Studies*, vol. 10, no. 4, pp. 357-365.

Schumpeter, J.A. (1926), *Theorie der wirtschaftlichen Entwicklung*, München, p. 134.

Schumpeter, J.A. (1942), *Capitalism, Socialism and Democracy*, New York.

Scott, W.A. (1968), 'Attitude Measurement', in G. Lindzey and A. Aronson (eds), *Handbook of Social Psychology*, vol. 2, Reading MA, pp. 204-273.

Sears, R.R. (1961), 'Transcultural Variables and Conceptual Equivalence', in B. Kaplan (ed), *Studying Personality Cross-Culturally*, New York, pp. 445-455.

Sexton, D.L. and Bowman, N. (1985), 'The Entrepreneur: A Capable Executive and More', *Journal of Business Venturing*, no. 1, pp. 129-140.

Shapero, A. (1975), 'The Displaced, Uncomfortable Entrepreneur', *Psychology Today*, vol. 9, no. 11, pp. 83-88.

Shapero, A. and Sokol, L. (1982), 'The Social Dimensions of Entrepreneurship', in C.A. Kent, D.L. Sexton, and K.H. Vesper (eds), *Encyclopedia of Entrepreneur-ship*, Englewood Cliffs NJ, pp. 72-90.

Sherif, M. (1947), 'Group Influences upon the Formation of Norms and Attitudes', in T.M. Newcomb, and E.L. Hartley (eds), *Readings in Social Psychology*, New York, pp. 77-90.

Sherif, M. and Harvey, O.J. (1952), 'A Study in Ego Functioning: Elimination of Stable Anchorages in Individual and Group Situations', *Sociometry*, vol. 15, pp. 272-305.

Siegel, A.E. and Siegel, S. (1968), 'Reference Groups, Membership Groups, and Attitude Change', in D. Cartwright, and A. Zander (eds), *Group Dynamics. Research and Theory*, New York, pp. 74ff.

Simmie, J. (ed) (1997), *Innovation, Networks, and Learning Regions*, London.

Simon, H(erbert)A. and Stedry, A.C. (1969), 'Psychology and Economics', in G. Lindzey, and A . Aronson (eds): Handbook of Social Psychology. (V): *Applied Social Psychology*, Reading MA, pp. 269-314.

Simon, H(erbert)A. (1982), *Models of Bounded Rationality; Behavioral Economics and Business Organisation*, Cambridge MA

Simon, H(erbert)A. (1957), *Models of Man*, New York.

Simon, H(ermann) (1996), 'Erfolgsstrategien unbekannter Weltmarktführer', Aus Politik und Zeitgeschichte, Beilage zu *Das Parlament* (B 23), April 31, 1996, pp. 3-13.

Singer, M. (1961), 'A Survey of Culture and Personality. Theory and Research', in B. Kaplan (ed), *Studying Personality Cross-Culturally*, New York, pp. 9-90.

Smilor, R.W. and Feeser, H.R. (1991), 'Chaos and the Entrepreneurial Process: Patterns and Policy Implications for Technology Entrepreneurship', *Journal of Business Venturing*, vol. 6, pp. 165-172.

Smilor, R.W. and Gill, M.D. (1986), *The New Business Incubator*, Lexington MA.

Smith, N.R., McCain, G. and Warren, A. (1982), 'Women Entrepreneurs Really are Different: A Comparison of Constructed Ideal Types of Male and Female Entrepreneurs', in K.H. Vesper (ed), *Frontiers of Entrepreneurship Research*, Wellesley MA, pp. 68-82.

Smith, V.L. (1994), 'Economics in the Laboratory', *Journal of Economic Perspectives*, vol. 8, no. 1, pp. 113-131.

Soete, L. (1985), 'International Diffusion of Technology, Industrial Development and Technological Leapfrogging', *World Development*, vol 13, no. 3, pp. 409-422.

Solow, R.M. (1994), 'Perspectives on Growth Theory', *Journal of Economic Perspectives*, vol. 8, no. 1, pp. 45-54.

Spilling, O.R. (1991), 'Entrepreneurship in a Cultural Perspective', *Entrepreneurship and Regional Development*, no. 3, pp. 33-48.

Stanworth, M.J.K. and Curran, J. (1976), 'Growth and the Small Firm - An Alternative View', *Journal of Management Studies*, vol. 13, pp. 95-110.

Stearns, T.M., Carter, N.M., Reynolds, P.D. and Williams, M.L. (1995), 'New Firm Survival: Industry, Strategy, and Location', *Journal of Business Venturing*, vol. 10, pp. 23-42.

Steinnes, D.N. (1984), 'Business Climate, Tax Incentives, and Regional Economic Development', Growth and Change, April, pp. 38-47.

Stevenson, H.H., Roberts, M.J. and Grousbeck, H.I. (1989), *New Business Ventures and the Entrepreneur*, Boston.

Stinchcombe, A.L. (1965), 'Social Structure and Organizations', in J.G. March (ed), *Handbook of Organizations*, Chicago, pp. 142-193.

Stöhr, W. (1986), 'Territorial Innovation Complexes', dans P.H. Aydalot (éd), *Milieux Innoateurs en Europe*, GREMI-Paris, pp. 29-55.

Stöhr, W. (1990), 'Synthesis', in W.B. Stöhr (ed), *Global Challenge and Local Response*, London, pp. 1-19.

Storey, D. (1986), 'Entrepreneurship and the New Firm', in J. Curran, J. Stanworth and D. Watkins (eds), The Survival of the Small Firm (I): *The Economics of Survival and Entrepreneurship*, Aldershot, pp. 81-101.

Storey, D.J. (1991), 'The Birth of New Firms. Does Unemployment Matter? A Review of the Evidence', *Small Business Economics*, no. 3, pp. 167-178.

Stotiand, E. (1959), 'Peer Groups and Reactions to Power Figures', in D. Cart-wright (ed), *Studies in Social Power*, Ann Arbor, pp. 53ff.

Strange, S. (1986), *Casino Capitalism*, Oxford.

Supiot, A. (1996), 'Perspectives on Work: Introduction', *International Labour Review*, vol. 135, no. 6, pp. 603-614.

Supiot, A. (1996), 'Work and the Public/Private Dichotomy', *International Labour Review*, vol. 135, no. 6, pp. 653-663.

Sweeney, G.P. (1987), *Innovation, Entrepreneurs and Regional Development*, London.

Tajfel, H. (1969), 'Social and Cultural Factors in Perception', in G. Lindzey and A. Aronson (eds), Handbook of Social Psychology. (III): *The Individual in a Social Context*, Reading MA, pp. 315-394.

Tarde, G. (1985), *Les Lois de l'Imitation*, Paris.

Thorngren, B. (1970), 'How Do Contact Systems Affect Regional Development?', *Environment and Planning*, vol. 2, pp. 409-427.

Tichonowa, N. (1997), 'Untersuchungen zum Kleinunternehmertum in Rußland', *Orientierungen zur Wirtschafts- und Gesellschaftspolitik*, vol. 74, no. 4, pp. 40-42.

Tissen, G. (1994), 'Die Reform der gemeinsamen Agrarpolitik und die EU-Strukturpolitik zur Entwicklung ländlicher Räume', *WSI-Mitteilungen*.

Törnquist, G. (1970), *Contact Systems and Regional Development*, Dept. of Geography, Lund.

Tushman, M.L. and Anderson, Ph. (1986), 'Technological Discontinuities and Organizational Environments', *Administrative Science Quarterly*, vol. 31, pp. 439-465.

Ullman, A.D. (1965), *Sociocultural Foundations of Personality*, Boston.

Ven, A.H. van de, Hudson, R. and Schroeder, D.M. (1984), 'Designing New Business Startups: Entrepreneurial, Organizational, and Ecological Considerations', *Journal of Management*, vol. 10, no. 1, pp. 87-107.

Vesper, K.H. (1983), *Entrepreneurship and National Policy*, Carnegie-Mellon Univ., Pittsburgh.

Waits, M.J. and Heffernon, R. (1994), 'Forging Good Policy on Business Incentives', *Economic Development Review*, vol. 12, pp. 21-24.

Wallace, A.F.C. (1952), 'Individual Differences and Cultural Uniformities', *American Sociological Review*, vol. 17, pp. 747-750.

Wallach, M.A. and Kogan, N. (1959), 'Sex Differences and Judgment Processes', *Journal of Personality*, vol. 27, pp. 555-564.

Wallach, M.A. and Kogan, N. (1961), 'Aspects of Judgment and Decision Making: Interrelationships and Changes with Age', *Behavioral Science*, no. 6, pp. 23-36.

Wallach, M.A., Kogan, N. and Bem, D.J. (1968), 'Group Influence on Individual Risk-taking', in D. Cartwright and A. Zander (eds), *Group Dynamics. Research and Theory*, New York, pp. 430-443.

Watkins, J.M. and Watkins, D.S. (1986), 'The Female Entrepreneur: Her Background and Determinants of Business Choice - Some British Data', in J. Curran, J. Stanworth and D. Watkins (eds), The Survival of the Small Firm (I): *The Economics of Survival and Entrepreneurship*, Aldershot, pp. 220-232.

Webber, M.J. and Joseph, A.E. (1977), 'On the Separation of Market Size and Information Availability in Empirical Studies of Diffusion Processes', *Geographical Analysis*, vol. 9, no. 4, pp. 403-409.

Weber, M. (1947), *Theory of Social and Economic Organization*, New York, (Parsons, T. (ed.)).

Webster, F.A. (1977), 'Entrepreneurs and Ventures: An Attempt at Classification and Clarification', *Academy of Management Review*, vol. 2, no. 1, pp. 54-61.

Weiss, W. (1969), 'Effects of the Mass Media of Communication', in G. Lindzey and A. Aronson (eds), Handbook of Social Psychology. (V): *Applied Social Psychology*, Reading MA, pp. 77ff.

Whiting, J.W.M. (1961), 'Socialization Process and Personality', in F.L.K. Hsu (ed), *Psychological Anthropology*, Homewood IL, pp. 355ff.

Whiting, J.W.M. (1968), 'Methods and Problems in Cross-cultural Research', in G. Lindzey and A. Aronson (eds), *Handbook of Social Psychology* (II), Reading MA, pp. 693-728.

Wiese, L. v. (1964), Definition von 'Organisation' in Handwörterbuch der Sozialwissenschaften (8) Stuttgart, p. 108.

Williamson, O.E. (1981), 'The Economics of Organisation: Transaction Cost Approach', *American Journal of Sociology*, vol. 87, no. 3, pp. 548-577.

Williamson, O.E. (1985), *Economic Institutions of Capitalism*, New York.

Wit, G. de (1973), 'Models of Self-employment in a Competitive Market', *Journal of Economic Surveys*, vol. 7, no. 4, pp. 367-397.

Wolfe, D.M. (1959), 'Power and Authority in the Family', in D. Cartwright (ed), *Studies in Social Power*, Ann Arbor, pp. 99-117.ff.

WZB-Mitteilungen (70), *Beliefs in Government*, pp. 15-28.

Zajonc, R.B. (1968), 'Social Facilitation', in D. Cartwright and A. Zander (eds), *Group Dynamics. Research and Theory*, New York, pp. 63ff.

Zimet, D. and Lachter, D. (1998), 'Industrial Parks as an Economic Development Tool in the Central Florida Panhandle', Paper presented at 37th Annual Meeting of the Southern Regional Science Association in Savannah GA on April 3, 1998.